Pierre Marie, J. D Souza-Leite

Essays on Acromegaly

Pierre Marie, J. D Souza-Leite

Essays on Acromegaly

ISBN/EAN: 9783744776875

Printed in Europe, USA, Canada, Australia, Japan

Cover: Foto ©berggeist007 / pixelio.de

More available books at **www.hansebooks.com**

THE NEW SYDENHAM

SOCIETY.

INSTITUTED MDCCCLVIII.

VOLUME CXXXVII.

ESSAYS

ON

ACROMEGALY

BY

DR. PIERRE MARIE

AND

DR. SOUZA-LEITE.

WITH BIBLIOGRAPHY AND APPENDIX OF CASES BY OTHER AUTHORS.

LONDON:
THE NEW SYDENHAM SOCIETY.

1891

EDITOR'S PREFACE.

A FEW words of explanation are required in reference to this little volume. It contains, in the first place (pp. 1 to 28), the original essay in which Dr. Pierre Marie, in 1885, announced to the profession his important clinical observations concerning a group of associated symptoms, of which the most conspicuous was overgrowth of the hands and feet and of the head and face. To this disease, with which he described a variety of less important concomitants, he gave the name of ACRO-MEGALIE. It has subsequently been suggested by others that in honour to its discoverer it should be known as *Marie's Malady*, but the name originally proposed is so descriptive, and has been so promptly and widely accepted, that it will probably not be displaced. As a very frequent accompaniment, if not as a cause, Marie observed the occurrence of great hypertrophy of the pituitary body. Marie's essay comprised not only his own cases, but had also references to a number of others which, unrecognised as a special disease, he had succeeded in finding already recorded in medical literature.

In 1890, six years after the publication of Marie's original essay, the subject having meanwhile excited the attention of the profession in all parts of the world, Dr. Souza-Leite, his pupil and friend, undertook the collection of such evidence as had accumulated respecting it, and published the valuable monograph which constitutes the second and larger part (pp. 30 to 160) of this volume. Part of what was published by the latter author was, however, a repetition of facts and details which had been given in the original essay. This circumstance has added somewhat to the translator's difficulties, since clearly there could be no object served by publishing the same cases twice in the same volume. It should be stated, also, that the original plan of the present editor was to prepare a full abstract of Marie's writings rather than a complete translation, and that it was only subsequent to its completion, for his own purposes, that his work was adopted for publication by the New Sydenham Society. It must be understood, therefore, by the reader that in many places the cases and statements both of Marie and of Souza-Leite have been considerably abbreviated. The editor trusts that he has in this way avoided diffuseness, whilst at the same time he hopes that he has not omitted anything of importance.

For the woodcuts, &c., which illustrate the essays, the Society is indebted to the kindness of Dr. Marie, by whom they have been supplied. Some of them, it is feared, exhibit the injurious effects of previous service, but it was not thought worth while on the part of the Society to undertake the cost of their reproduction, nor, indeed, were the original drawings at its disposal.

At the conclusion of the second essay, page 81, a list of all the cases recorded both by Marie and Souza-Leite is given, and to this is added a number of additional cases collected from various authors who have written since.

P. S. H.

12, OLD CAVENDISH STREET;
October, 1891.

CONTENTS.

	PAGES
Two Cases of Acromegaly; an unusual Non-congenital Hypertrophy of the Head and Upper and Lower Extremities. By Pierre Marie	1—28
A Thesis on Acromegaly—Marie's Malady. By Souza-Leite	31—160
Bibliography of Acromegaly. List of Cases and Additional Cases. By the Editor	161—180
Index	181

TWO CASES OF ACROMEGALY;

AN UNUSUAL NON-CONGENITAL HYPERTROPHY OF THE HEAD AND UPPER AND LOWER EXTREMITIES.

BY

PIERRE MARIE, M.D.,

DIRECTOR OF THE LABORATORY ATTACHED TO THE SALPÊTRIÈRE.

A Thesis republished in the 'Revue de Médecine,' 1885.

TRANSLATED BY

PROCTER S. HUTCHINSON, M.R.C.S.,

ASSISTANT SURGEON TO THE HOSPITAL FOR DISEASES OF THE THROAT.

TWO CASES OF ACROMEGALY.

The two cases which are the subject of this article were observed in Professor Charcot's clinic. They present a disease which has not yet been separately described in all its features. Yet it seems to possess a special morbid entity, for among all the patients are found the same symptoms and characters. Whether this disease is very rare, it is difficult to say, as so little has hitherto been known concerning it. We have simultaneously observed two cases, but though we have made extensive literary researches, we have been able to find on record only a very small number of examples of it.

Case I. Fusch, a woman, aged 37, single.—There was no family history of rheumatism or nervous disease. Her mother and father had died of chest disease; the rest of her relatives were healthy. The patient herself had always been strong and healthy, and had never had rheumatism or syphilis. At the age of twenty-four, following great fatigue and exposure in washing a house, her menstruation suddenly ceased. Fifteen days afterwards she took to her bed, with shivering, general weakness, and trembling in all her limbs. At the end of four or five days her menses reappeared. She remained in bed three weeks; there was no pain in the joints, but she noticed, on raising herself, a weakness of the left hand, with tingling sensations in it. At the end of some weeks the strength returned to a great extent in this hand, though not completely. From this time she went for three months without menstruating.

Having again been exposed to draughts, she was seized with headaches, and pain in the back and arms. There was no redness nor swelling of the joints. She was not laid up, but noticed that there were grating sensations in the shoulders and knees. She took sitz baths and the menses reappeared, but have since not returned. She experienced a sensation of sinking at the pit of the stomach, which seemed to be empty a quarter of an hour after food, when she had a fresh desire to eat. She was treated by a doctor for anæmia. After resting for a time, her strength came back, and she was able to recommence work, but was not so strong as formerly. She had now for seven years lived in the country, and done very heavy work. In 1880 she came to Paris, being then aged thirty, and remained in a somewhat more satisfactory condition till the commencement of 1883. After this she was troubled with violent pains, sometimes over the forehead, the parietal eminences, or the temples. These pains often prevented sleep, and they had been worse lately. The patient had always had somewhat large limbs, but nothing compared to their present size. It was at the age of twenty-four, at the time the menstruation suddenly ceased, that she noticed the sudden increase in her hands. Her face at this time also underwent changes, which will be referred to later, so that when the patient returned home none of her relatives could recognise her.

Present condition.—The whole feet are large, including the toes. Though the latter are increased in size, they have preserved their form, there is no true deformity, their appearance is simply that of a very big person.[1] The nails, skin, &c., show no change. The tibia is not increased in size, but there is a marked projection at the inner side of the knee, which, as there is no effusion into the joint, is due either to thickening of the patella or of the inner condyle of the femur. There is grating felt on moving the knee-joint, but not so marked as in dry arthritis, and due possibly to friction of the fibrous tissues.[2] All the movements of the leg and foot are normal. The chest shows nothing peculiar, beyond a marked posterior curve in the dorsal region. This, though

[1] Detailed measurements of the foot and leg are given in the original.—Tr.
[2] Detailed measurements of the thigh are given in the original.—Tr.

not angular, is distinctly marked. The hands are very large, but of regular form; their thickness and width are relatively greater than their length, and attention is at once attracted to them on seeing the patient.[1] The joints of the fingers

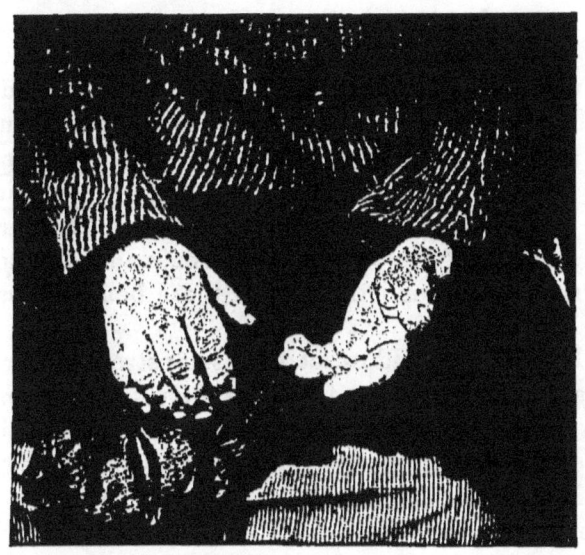

Fig. 1.—The hands are enormous, with very marked pads. Photograph taken in 1885.

are not specially enlarged in proportion to the bones, which have certainly shared in the hypertrophy; the fingers present a somewhat flattened appearance. The width of the nails is increased. There is apparent atrophy of the interosseous muscles, which, according to the account of the patient, took place early in the disease. Movements of the fingers are well performed. On making passive movements of the shoulder-joint, very marked grating is felt. The muscular movements of the left arm and neck seem weaker than normal. The thyroid cartilage seems increased in size, but the thyroid gland is found with difficulty, and is probably less developed than normal. The tongue is enlarged. The patient is a little deaf, and the sight is also slightly defective.[2] There are two symmetrical depressions

[1] Detailed measurements of the hands are given.—Tr.
[2] Detailed measurements of the head are given.—Tr.

in the parietal bones, that on the left side being less marked than on the right; it being over the latter that the patient complains of intense persistent pain. Briefly, the face presents the appearance of a lengthened ellipse, with the diameter from above downwards. The centre of this ellipse is situated on a level with the bridge of the nose, and its greatest diameter is opposite the malar prominences. The cranial vertex is of nearly the same size as the end of the chin. The lower jaw is well developed. The complexion is pale, and the eyelids a little pigmented. The patient's thirst is intense, obliging her to beg tea of her friends in order to satisfy it. The quantity of urine is excessive; but no measurement of the amount has been made, and no record of analysis kept beyond the fact of there being no sugar.

CASE II. Heron, a widow, æt. 54, a sempstress.—Her mother died in a confinement, her father at fifty-six from consumption; grandparents unknown. Five brothers also died consumptive; the sixth, aged seventy, suffers from rheumatism. The patient herself had no illness up to the age of fifteen, when her first attack of rheumatism (?) occurred. Beyond this and an attack of smallpox she had good health up to the age of twenty-nine. She was regular from the age of thirteen. She was married at sixteen; one child. At twenty-nine, the date of commencement of the illness, she had a fall from a height on a heap of snow, which caused considerable nervous shock. Her menses disappeared immediately, and have not since returned. During the first few months which followed she fancied herself pregnant, and did not take much notice of it. Three months after the fall she complained of general cramps, and had to take to bed for about eight days. Two months later, that is, five months after the accident, abscesses appeared in several parts, in the axillæ, under the jaw, and in the labia majora. At this time the patient noticed that her waist had increased, but as yet her hands and feet had not changed. At thirty she began to suffer from hæmorrhoids, with frequent loss of blood, which occurred regularly every three months. The patient was still strong and could walk about, but became suddenly blind, remaining so for eight months. At the end of this time she recovered sight partly, so as to be able to thread a needle, but the improvement was only transitory, and two years later she lost her sight permanently. Extensive corneal opacities prevented examination of the fundus. About the age of thirty-two the patient became very weak, and could not walk five minutes without being obliged to sit down. It was also about the same period that she noticed, in trying to put on a last year's hat, that her head had increased in size. The patient then began to grow very rapidly. She had previously been known as "la petite," and she had worn boots with high heels in order to increase her height, her short and slim figure being remarked on. Her dresses had now to be increased in length, and the circumference of her waist was found to be enlarged. At this time also her hands and feet began to progressively enlarge. It is necessary, however, to add that in recent years the latter have decreased somewhat, the patient stating definitely that they were once more hypertrophied than they are now. The feebleness from which the patient suffered lasted about a year; since then the strength had returned little by little, and she was again able to walk. At thirty-three a second attack of rheumatism occurred, which lasted about six months. The patient could walk, but could not get upstairs.

In 1869 she was admitted into the Salpêtrière. She still suffered from hæmorrhoids.

In 1872 she had violent neuralgia, and nearly all her teeth fell out.

Present condition.—The fingers show very marked nodosities, particularly the joint between the first and second phalanges. They are more marked on the dorsal aspect and sides than on the palmar surface. That of the middle finger is most definite. There is possibly some atrophy of the interossei. The nails show no change beyond longitudinal striation. The spine presents a lateral curvature in the dorsal region, and also an anterior curvature in the cervical region, which makes the patient slightly round-shouldered.[1] The ribs project more at their anterior angle, and are distinctly enlarged. The thickness of the pelvic bones seems a little increased, particularly near the anterior superior iliac

[1] Dimensions of clavicle are given, showing some increase.—TR.

spine. At the age of eighteen the patient wore a boot of size No. 37; it is now necessary to wear No. 48 or 50. Both

FIG. 2.—The hands are somewhat large, and especially nodular.

feet are considerably increased in size; the toes are very much larger than those of an ordinary person, especially the great toe, the end of which is a little clubbed. There is no alteration in the bones of the limbs, the knees only being thickened. The muscles of the calf are considerably diminished in size. On the dorsum of the feet, and on the front of the leg, the skin is pigmented, and varicose veins are present in the leg and thigh. The face presents a long oval. The borders of the orbits are very thick, also the frontal eminences, making between them and the upper border of the malar bone a deep depression, something similar to the corresponding region of a cow. The nose is large. The lower jaw is very thick, and increased in all its dimensions. The lower lip is more prominent than the upper. All the sutures of the skull, but especially those between the frontal and parietal, and parietal and occipital, show a marked crest-like prominence. At the upper angle of the occipital bone,

Fig. 3.—The face is lengthened, the forehead low, and the chin much thickened.

at its point of junction with the sagittal suture, is a considerable thickening. The temporal bones also project from the surface of the cranium. The tongue is enlarged and fissured, but moist. The movements of the arms seem rather weak, also the extensors of the leg.[1] The thyroid gland can scarcely be felt, and is certainly diminished. The laryngeal cartilages are possibly a little large, though not definitely hypertrophied. The ears are decidedly enlarged.[2] The general look of the patient was a little cachectic, and she was obliged to take to bed on account of general weakness (and also for the hæmorrhoids). During the first few years of her illness she suffered from excessive thirst, drinking one after another

[1] Detailed tests of the muscular movements are given.—Tr.
[2] Detailed measurements of the skull and limbs are given.—Tr.

four or five pints of water; for a long time, however, this thirst has not been present, except very occasionally. Examination of the urine showed nothing abnormal.

As has been seen, in these two cases the principal feature which first attracts attention is a considerable increase in the size of the hands, the feet, and the head; besides these there are some other symptoms to which we will refer later. First as to the changes in the extremities. In Case I, the hands particularly were enlarged, and to such extent, that when the patient came to Professor Charcot's out-patient department, though she complained of a violent neuralgia in the head, it was these that first attracted attention. The entire hand had undergone an increase in size, more perhaps in width than in length, which give it a decidedly flattened appearance. Their form, however, as a whole was not altered, nor was there any deformity properly speaking; if they belonged to a giant their appearance would present nothing abnormal.[1] The fingers are also very large, rounded at their ends, and throughout their length of about equal size.

In Case II, the hand is distinctly less thickened than in Case I, but it appears on the statement of the patient that it had once been larger. However that may be, it is certain that they are larger than normal. The fingers, especially at the level of the inter-phalangeal articulations, show marked nodosities, which were not present in Case I. The nails were enlarged, a little flattened and striated, but without marked change. In contrast with this considerable hypertrophy of the hands, it is peculiar to see the forearm and arm preserve their ordinary size; and especially to note that while the skeleton of the hand has assumed the proportions of a giant, the radius and ulna show no increase in their dimensions. We shall note later, especially in the cases of Friedreich, that these bones may also undergo increase in size, but to a much less extent than the hands or feet. In regard to the feet, all that has been said concerning the hands is applicable. They have undergone considerable hypertrophy, without appreciable deformity. Their bones

[1] We remember to have seen a giantess who went about under the name of the "Chinese Giantess," whose hands had very similar proportions to those present in our patient; she was simply taller.

are also enlarged, while the tibia and fibula are normal. We have noted that in Case II the feet were comparatively more hypertrophied than the hands; in Case I the reverse was present. As to the soft parts of the hands and feet, they show no structural change, beyond having undergone proportionate development with the bones.

Besides this hypertrophy in the extremities, the head shows an increase in some of its parts; it is chiefly in the bones of the face that this is present; most of them we have found hypertrophied, but more particularly the bones of the nose, the malar bones, and the lower jaw. The latter in both our patients has undergone such increase that it projects beyond the upper jaw, being enlarged in all its dimensions. The general contour of the face is that of an ellipse, with the vertical diameter longest.

The cranium has also suffered; the frontal eminences are slightly thickened, which with the hypertrophy of the malar bones exaggerates the depression formed by the anterior part of the temporal fossa. As regards the other regions of the skull, the changes are not alike in our two cases. In Case I we have noted the presence of two marked depressions in the parietals. In Case II, however, on the contrary, projections are found in the form of crests close to the sutures. In general the dimensions of the cranium appear a little increased, and in Case II the patient mentions among other things that at one time her bonnets became too small; but what increase there is in the size of the cranium is not to be compared with that of the bones of the face. Such are the principal characters which at first sight attract our attention, and give a special aspect to these patients. They are not, however, the only ones. To refer again to the changes in the skeleton, we must note the anterior curvature of the spine which was specially marked in Case I; in consequence of which the patient had much difficulty in raising the head, and usually kept the chin nearly in contact with the sternum. It must not be forgotten also that other bones may be more or less enlarged, as, for example, in Case II, the clavicles, the ribs, the patella, and the ilium. But what should be specially noted is the almost complete exemption of the long bones of the upper and lower extremities.

As regards the fibro-cartilages, they seem to have a tendency to hypertrophy; as in Case II it was observed in connection with the ears, and possibly also with the larynx.

The joints, on the contrary, for the most part were not affected; grating was, however, present in the shoulder-joint of Case I, and nodosities on the phalanges of Case II. Possibly also lesions of the vertebral joints had caused the spinal curvatures, at least the latter were not due to hypertrophy of these bones. But at any rate it may be noted that there were no ankyloses, deformities, or exostoses in the neighbourhood of the joints. Conversely to what is present in the skeleton, the muscles are diminished in size, though it is difficult to decide whether this is due to true atrophy or to want of development. Besides diminution in size, considerable weakness of muscles was also present, as has been shown in some detail in the narratives of the cases.

As regards the organs of special sense, they present more or less definite changes. In Case I there was a little deafness; in Case II complete blindness, of which we do not know the cause beyond what has been stated; *i. e.* an old irido-keratitis, with extensive corneal opacity. While referring to the organs of special sense, we must note also the manifest hypertrophy of the tongue, which definitely existed in both our cases. The general sensibility has, however, been found intact throughout, but there were occasionally very violent pains, which had been present for a certain time in Case II, and had then disappeared. In Case I, on the contrary, they had been present almost since the beginning of the illness; and had been situated in the head, the back, and the arms. During the last few months they had become more intense, and occupied almost entirely the fore and upper part of the head, presenting exactly the features of facial neuralgia. It was solely for these neuralgic pains that the patient came to the out-patients' department of Salpêtrière; she did not say anything concerning her deformity, and it was only when she carried her hands to her face that their extraordinary size was noticed.

With regard to the circulation and blood-vessels there is only one thing to note, the tendency to varicose veins which existed especially in Case II. In this patient, in fact, there were not only very distinct venous dilatations with varices, but also large hæmorrhoids, which had caused much suffering, and very abundant loss of blood. We may add that our two patients had the thyroid gland diminished in size, at least as far as could be judged by palpation.

As regards the skin, it has been seen to be absolutely healthy; showing no structural peculiarity beyond a slight degree of pigmentation of the legs in Case II. In their general health these two patients showed a certain degree of cachexia, not strictly speaking indicative of a grave disease, but rather of a general weakness. It may, however, be noted that during the many years that they have suffered from their complaint nothing has indicated an approaching fatal termination; the disease is essentially chronic.

The chief organic functions appear to be performed normally; we have only to note in Case I a very intense thirst, with abundant urine; in Case II a similar condition also existed.

On the question of etiology we have nothing definite. In both cases the patients attributed it a great deal to chills, and in Case II a fall was mentioned, but it is well known what importance is usually to be attached to these two causes, cited so often for such different maladies. In Case II there had been two attacks of rheumatism. It is perhaps to this cause that the nodosities on the joints of the fingers should be attributed. On the other hand, in Case I the patient stated that she had never had rheumatism.

One fact, however, which seems to present a certain amount of importance, if not from the etiological point of view, at least from the question of symptoms, is the disappearance of menstruation coincident with the commencement of the illness. In both cases this was expressly mentioned.

As to age, it may, we think, be safely asserted that it is a malady of adult life, coming on from fifteen to thirty-five. This statement is based upon the record of other cases which we shall bring before the reader. These cases, though briefly recorded, offer, we assert, very striking resemblances to the two which we have described above. It will be noticed that they concern men, whereas our two patients were women; so that both sexes may be affected by the disease.

Cases by other Authors.

The five cases which follow, published by different authors, under different names, ought, we believe, to be considered as examples of the same disease; for the similarity of symptoms in these and in our own patients amounts almost to identity.

Case III (Saucerotte-Noël).—He has been an inhabitant of Mangoville for six years, æt. 39, height five feet two to three inches, of thin and slender stature. Up to the above age, he had for about six years noticed that all the bones of his body were enlarging little by little, without lengthening, to such an extent that they became double their former size. He presents most unusual features, both in general proportions and in the separate limbs. His bones appear to have become thickened at the expense of the muscles, which are flaccid and atrophied. The man is obliged to have his hats made for him, not finding any ordinary shape sufficiently large. His eyeballs are level with the rest of his face, owing to the thickening of the bones of the orbit. His lower jaw has lengthened more than the upper, probably on account of its being a separate and moveable bone; resulting from this the lower incisors project in front of the upper to the width of a finger. The lower lip is perhaps the only soft structure which has followed the progressive enlargement of the bones, it being very large. The vertebral column is of increased size; and the same with the clavicles, scapulas, and innominate bones, which have extended enormously in length and thickness. The ribs and sternum are also increased, so that the chest has become prominent and the stomach flat; atrophy of the muscles also augmenting the difference. The ribs are about one and a half inches in width, and here and there border on one another. The feet and hands may be compared to those of an ordinary person with very large limbs. The legs appear at first sight to be out of proportion with the rest of the body, but this in reality is not so, as they consist mainly of bone, having hardly any soft structures, though the tendo Achillis itself is nearly twice its normal size.

This individual does not attribute the growth of his bones to any illness; he has always been a great eater, but has also taken much exercise, being one of the great farmers of the province. He is now almost always sleepy, no doubt owing to the thickening of the skull compressing the brain. For about two years he has experienced oppression of the chest, no doubt in the same way due to embarrassment of the lungs. It may be added that since the increase has reached this point, his pulse is often so small that it cannot be felt. The patient has taken all sorts of medicines in the hopes of diverting nutrition from the bones. The question may be asked, why the blood should have left the easily distended vessels to flow in the less dilatable ones? He experiences, I forgot to mention, general pains which he attributes to rheumatism, but it is more likely that they are the result of distension of the periosteum.

In 1766 the patient weighed 119 lbs., and last year 178 lbs., atrophy of the muscles at the same time having taken place.[1] The lower jaw is eighteen inches in length from one condyle to the other, and four inches in depth at the incisor teeth, which makes it rest nearly on the sternum, and gives the patient the appearance of having no neck. The movements of the joints are rather painful, especially those of the fingers and toes. Examination of the urine showed absence of the ordinary salts, which Saucerotte suggests had been carried to the bones.

During 1771 the further increase in the bones was noticed. In May, 1773, improvement in the breathing, &c., had taken place; but in July, 1773, after living too freely, his abdomen had become tense and hard, with extreme nausea and thirst. Purgatives were ordered, but the breathing became worse, the head heavy; general swelling set in, and a tumour formed in the neck of a soft and emphysematous nature. Finally, after taking an overdose of purgatives, inflammation of the lower intestine set in, hæmorrhoids formed, and

[1] Measurements of the head given.—TR.

the patient after a few months died. No autopsy was allowed, but the bones were finally dug up by Saucerotte, and a rib, sternum, and clavicle sent to the Dupuytren Museum.[1]

It appears as if probably the following case recorded by Alibert belongs to the same category.

CASE IV (Alibert).—"I brought to my clinical lectures a short while ago Pierre, of * * *, æt. 32. He had been puny at birth, and had remained small till puberty, at which period he suddenly grew to the height of six feet four inches, his limbs becoming a proportionate size. His face was oblong, and his tongue of considerable width. His voice was harsh, and resembled that of an old man. He experienced aching in his legs over the loins. He was tormented with such an intense thirst, that he drank up to eighteen bottles of pure water every day; and his urine was proportionately abundant. Besides other symptoms there was want of sexual power."

CASE V (Friedreich, *Résumé*).—Wilhelm Hagner, 26 years old, a shoemaker. Among six brothers and sisters, one brother has suffered from a similar affection. Nothing noteworthy as regards previous illnesses. Without any definite cause, the patient noticed at the age of eighteen that the feet, particularly in the neighbourhood of the phalanges, were becoming progressively larger. The legs were then involved, and the same with the knees. In walking "his legs felt as if made of lead." About two years later both hands became larger, the fingers became markedly thickened, and, from the tension produced, work became impossible. Subsequently the fingers became less swollen, and he resumed work. During the last two years the disease has not progressed. He has experienced no pain.

At the present time the hands, feet, and legs present the appearance of elephantiasis; but pressure shows that the increase is due to thickening of the bones. The phalanges of the fingers and toes, and also of the bones of the hands

[1] Details of measurements of these bones given, showing enlargement specially of the ends of the clavicles.—TR.

and feet, are enormously thickened, and certainly a little lengthened. In the legs and forearms the growth is specially marked at the epiphyses; the shafts themselves, however, are also enlarged. The borders of the tibiæ and fibulæ, the bones of the tarsus and carpus, as well as the patellæ, are all affected with hypertrophy. Everywhere the bones have a smooth contour, nowhere are there exostoses. The bones of the thigh and of the arm proper are evidently hypertrophied, but relatively to a less extent. Besides the bones of the extremities other bones of the skeleton are also hypertrophied. Thus the sternum is larger and more massive; also the crests of the scapulas and iliac bones are notably thickened. The ribs are very thick and wide, and at certain places leave only a narrow space between them. The spinous processes of the vertebræ, particularly the lower cervical and upper dorsal, are distinctly thickened. The clavicles are also about double their normal size. Among the bones of the face, the malar and specially the palate bones, as well as the alveolar process of the maxilla, are affected; whereas the teeth show no increase. The hyoid bone is remarkably large. The cranial vault shows no deformity.[1] Certain of the cartilages are also increased, as the tarsal cartilages, the epiglottis, and to a less extent the nasal septum. Other cartilages of the larynx are, however, not affected. The skin over the hands and feet appears somewhat thickened, but nowhere rigid. The nails are enormously increased, both on the feet and hands.[2] The muscles generally, especially on the extremities, are flabby and badly nourished. Standing and walking easily bring on fatigue, and the hands are weak. The patient prefers cold to heat. The cutaneous sensibility is intact, and the internal organs seem normal. The excretions are normally performed. The patient has, however, noticed a marked tendency for the feet to perspire since the commencement of the illness. Analysis of the urine shows nothing abnormal.

CASE VI (Friedreich, *Résumé*).—Carl H——, æt. 22, brother of the former patient. In him the first symptoms showed

[1] Detailed measurements of limbs, &c., here given.—TR.
[2] Measurements given.—TR.

themselves at the age of seventeen, that is, about the same time as in his brother. They commenced without definite cause, and without pain. The progressive enlargement invaded first the feet, then the legs and knees, so that "each time the shoemaker had to make his boots larger." About a year later the hands and forearms became affected; for the last two years it has not made any progress. At the time when Friedreich first examined him, in July, 1867, he showed the same enormous thickening of corresponding bones as in his brother. There was, to a slighter extent, hypertrophy of the femurs, humeri, ribs (particularly at their costal ends), the sternum, pelvis, scapulæ. The patellæ and clavicles, however, were hypertrophied to a greater extent. Among the bones of the face the malar bones were especially enlarged; the hypertrophy of the hard palate and alveolar process, present in his brother, was, however, not well marked. In Carl, as in Wilhelm, the same smoothness of the bones was found, and proportionate hypertrophy, especially of the epiphyses of the long bones. There was the same enlargement of the hands and feet, and also of the nails. As to the cartilages, there was simply thickening of the septum, the ears, eyelids, and epiglottis not being involved. In Carl the muscular system was in general stronger and better nourished than his brother, and he could do fairly heavy work. In him also the sensation of heat and tendency to perspire were absent. The internal organs and excretions were perfectly normal.

The following case, published in 1877 by Professor Henrot, seems to belong to the same category;[1] we will mention later our reasons for not holding the opinion of this distinguished physician, that it is a case of myxœdema.

CASE VII (Henrot, *Résumé*).—A man, æt. 36, no history of syphilis. Since the age of six there have been enlarged submaxillary glands. Since the age of fifteen he has noticed progressive enlargement of the feet and hands. When twenty he became a soldier for seven years.

[1] 'Notes de Clinique Médicale,' by Dr. Henri Henrot, Rheims, 1877 and 1882.

January, 1877.—The lower part of the face shows unusual development; there is no demarcation between it and the upper part of the chest. Both lips are swollen, the lower one being the thickness of a finger, and being turned down so as to leave the mouth half open. The tongue is at least double its normal size, and completely fills the mouth, projecting between the lips; it presents a normal appearance, and seems to be hypertrophied throughout. The upper dental arch is normal, but the lower one is much increased, so that when the jaws are closed the index finger can be inserted between the teeth. The teeth of the lower jaw show increased size in all their dimensions. The eyeballs project considerably, and the countenance lacks vivacity. The lids, though not œdematous, are thickened, the upper one covering half the cornea. The pupils are equal, moderately dilated, and act but little to light. The nose is excessively dilated. The width of the face is considerably increased. The whole circumference of the lower jaw is covered with glandular enlargements, forming one mass, and so hard and intimately connected with the bone itself, that it is difficult to distinguish between them. These tumours in the parotid and submaxillary regions are continuous without line of separation from the upper part of the chest, the hypertrophied thyroid being welded in the common mass. The forehead and cranium show no deformity, though at first sight, in comparison with the face, appearing atrophied. The complexion is a dull white. The upper and lower limbs are of normal size, and in contrast with them the hands and feet appear out of all proportion. Their contour is not altered, but they are uniformly hypertrophied in all their dimensions, and the nails are large and flattened. The movements of the arms and legs are weak, probably partly from the hypertrophy of the hands and feet, and also from muscular weakness. The patient lies helpless in bed. The cutaneous sensibility seems dulled, but there is no anæsthesia. His hearing is good, and he answers questions fairly intelligently. He complains of constant severe headache. His vision is very imperfect, and he experiences a difficulty in seizing objects.

Autopsy.—Skin healthy. The thyroid gland is increased to four or five times its normal size. The submaxillary glands are lost in immense glandular masses, which surround the lower jaw, the latter presenting rather the appearances of that of a huge ape. These glandular masses are of homogeneous structure, and show no tendency to degeneration.

The pneumogastric, glosso-pharyngeal, and brachial plexus have undergone a definite increase in size. The bones of the cranium are neither thickened nor deformed; the base of the skull being normal, except the sella tursica and the pituitary fossa, which are enlarged. On the lower jaw, on the hyoid bone, the tibia, fibula, radius, and ulna, in fact on all the bones which have been preserved, are well-marked osteophytes. On the lower jaw, which may be taken as an example, the mental spines form large projections the size of a haricot bean. On the hyoid bone the tubercles for the insertion of muscles form marked prominences, and there are also bosses on the tibia where the ligaments are attached. On the outer surface of the ascending ramus of the lower jaw are hard, eburnated areas, studded with a number of little holes.

The hypertrophy of the hands and feet is not confined to the soft structures, it affects also the bones, longitudinal section of them allowing no doubt on this point.

These osteophytes, which are developed wherever prominences normally exist, are more marked on the lower than the upper limbs. There is no alteration in the medullary canal or cancellous structure. The cartilages of the larynx (thyroid, cricoid, and arytænoid) are ossified, and their projecting points enlarged.

As regards the nervous system, Dr. Henrot found, besides bony plates in the spinal meninges, an ovoid tumour the size of a hen's egg at the base of the brain, situated in the middle line, in the position of the pituitary body. Its width measured 45 mm., and its length 30 mm. It was bounded by the following organs: in front, the commissure of the optic nerves, which was flattened and formed a large flat band; at the sides the sphenoidal lobes of the brain showed depressions; and behind it rested on the cerebral peduncles. Though mainly of regular form, it presented in front and to the left a small projection which seemed superadded to the rest. Its consistence was somewhat soft, the anterior part differing in appearance from the posterior. On carefully raising it, its close connection with the base of the brain was seen, extending from the infundibulum to the tuber cinereum. The infundibulum was more developed than normal, and the pineal gland was at least double its ordinary size.

There was present besides a considerable hypertrophy of all the ganglions and nerves forming the sympathetic.[1]

It will be seen that the symptoms in this patient are almost exactly those which were present in the two others which we have recorded, *i. e.* enlargement of the face, and hypertrophy of nose and eyelids, prominence of lower dental arch, and hypertrophy of lower jaw, considerable increase of hands and feet, muscular weakness, tongue hypertrophied, and intense thirst.

It is true that in some points this case differs a little from the others, but these seem to us rather variations in the course of the same malady than true differences. Such deviations may be instanced as the increase in the submaxillary glands, the absence of deformity of the bones of the skull, excepting the sella tursica; the presence of sugar in the urine, increase in the thyroid gland, and thickenings on the bones. More numerous points of resemblance than of divergence may be said to exist; and the former have more importance as regards the essential features of the disease than the latter. On the other hand, the differences from myxœdema may be said to be still greater, as seen in the hypertrophy of the skeleton, and notably of the hands and feet; the prominent lower jaw, the elongated face, the increased thyroid, which can none of them be said to be symptoms characteristic of myxœdema. Nor in the description of the skin by Professor Henrot is there anything similar to that characteristic of myxœdema. The latter disease is, we think, definitely excluded; its chief features, we repeat, being an alteration of the skin and subjacent tissues, and not a modification of the bones; the latter being, on the contrary, a characteristic feature of the disease in question.

A certain number of diseases, accompanied by increase in the size of the bones, might be confused to some extent with

[1] For details see report of Dr. Henrot. The microscopical examination was not completed at the time of writing.

the disease which we have described. Among them may be specially noted the overgrowth of the face and cranium, which Virchow[1] has described under the name of leontiasis ossea; and to which category belong the cases of Jadelot,[2] Forcade, probably also those of Murchison[3] and Adams.[4] The interesting case of that excellent authority, M. Le Dentu,[5] ought perhaps also to be classed among them, though its short duration makes it somewhat doubtful in which it should be placed. What distinguishes leontiasis ossea from the present disease is the want of hypertrophy of the hands and feet; also the development of exostoses on the facial bones and cranium, producing hideous deformities; whereas in our own cases the bones of the face and cranium are the seat of more uniform hyperostosis, without important tumours or bossy outgrowths.

A fairly large number of cases have been recorded in which hypertrophy of the face with or without hypertrophy of the limbs existed, but affecting only one half of the body. It has been met with more often as a congenital malformation, and the cases have been specially described by Trelat and Monod; and more recently in a monograph by Lewin,[6] in which are reviewed all the cases of atrophy or unilateral hypertrophy of the face, body, hands and feet, &c. We have carefully gone into all these cases, and feel confident that none of them comprise the malady which is made the subject of this work. The same also may be said of certain cases recorded by Fereol,[7] Wrany,[8] and Schutzenberger:[9] as regards the first we agree with Dr. Fereol that it was a case of rheumatism affecting various parts of the skeleton; with regard to the two other writers the details are insufficient to found a definite diagnosis.

[1] Virchow, 'Pathologie des tumeurs,' trad. Franc., t. ii, p. 22.
[2] Jadelot, 'Case of Cranial Tumour,' Paris, 1799; also Gervais, 'Journal de Zoologie,' 1875.
[3] Murchison, 'Path. Soc. Trans.,' vol. xvii, p. 243.
[4] Adams, 'Path. Soc. Trans.,' vol. xxii, p. 204.
[5] Le Dentu, 'Revue de Médecine,' 1879.
[6] Lewin, 'Charité Annalen,' ix Jahrgang, 1884.
[7] Fereol, 'Bulletin de la Société Clinique,' 1877, p. 99.
[8] Wrany, 'Prager Vierteljahrschrift,' 1867, p. 73.
[9] Schutzenberger, 'Gaz. Méd. de Strasbourg,' 1856, No. 4.

There is, however, a disease which may be confounded with that with which we are now occupied, for it also produces increase in the size of the head and limbs. This disease is Osteitis Deformans.

Perhaps it will not be out of the way to say here a few words on this peculiar malady, first described in 1876 by Sir James Paget.[1] This malady, which should by rights be called "Paget's disease," has been specially studied in England. Without giving a complete description we will briefly recall, from the works of the author above mentioned, what are its chief characters. First, however, we will give a rapid summary of the cases of this kind published in France under different titles. As far as we know, Dr. S. Pozzi is the only author who has described cases of this disease under its true designation. It will be noticed that the disease has already been observed in France for many years, and that it agrees exactly with the characters described by Paget.

CASE VIII (*Résumé*).[2] ——, house painter.—At thirty-five he had suffered from lead paralysis. At forty he perceived by his hat that his head had increased in size. At forty-four years he complained of weakness of the legs, with occasional swelling; and a little later of severe pains in the legs and shoulders. At fifty, the œdema having disappeared, increase in the size of the bones of the legs was noticed; his clavicles were also larger. At fifty-one Dr. Vulpian recorded hypertrophy of the bones of the head, clavicles, humerus, ulna, femurs, tibias, fibulas, bones of tarsus, ribs; and, less manifestly, of the iliac bones, sternum, and metatarsus. At this time he also had very severe pain in the shoulders, arms, and lower limbs, with giddiness, dimness of vision, and dulness of hearing. At sixty-six, Drs. Rathery and Leloir noted increase in the size of the head, with a crest at the parietal suture. There was thickening on the side of the cranium, specially at the orbital fossæ. The bones of the face appeared normal; but the horizontal ramus of the lower

---

[1] Paget, 'Med.-Chir. Trans.,' vol. lx, 1877; lxv, 1882.

[2] Drs. Bourceret, Rathery, Leloir, and L. Guinon, 'Revue de Médecine,' 1881. The autopsy of this case showed on microscopical examination changes similar to those described by Paget and Lunn.

jaw had increased in size, and the patient sometimes experienced sharp pain in it. Besides the hypertrophy already noticed, the thorax showed a spinal curvature, giving the patient a hump-backed appearance. The radius and the bones of the wrist and hand appeared healthy. The bones of the tarsus and metatarsus were slightly hypertrophied, the phalanges being healthy. The muscular system, except the pectorals and muscles of the leg, which were in some instances atrophied, were healthy. Intelligence feeble.

CASE IX (Binet, *Résumé*).[1] ———, a woman æt. 60.—At fifty-seven the disease began in the legs, the tibias of which are thickened, and their curve increased. In the forearms the anterior curve was increased and hypertrophy existed, specially of the ulna on the right and the radius on the left. No mention is made of increase in the size of the hands, feet, or head. "The rest of the skeleton appears normal." She had dull pains in the legs, and sometimes shooting; also weakness of the left arm; no other nerve symptoms or muscular atrophy. Reflexes normal. Eighteen months ago she had transitory left hemiplegia.

CASE X (Richardière Rogier, *Résumé*).[2] ———, a copper turner.—At forty-five he noticed a difficulty in moving the right forearm, and began to limp and to bend forwards. The disease progressively advanced, with hypertrophy and deformity of the head, bones of the limbs, thigh, humerus, and forearm. There was very little increase in the size of the phalanges of the fingers, or of the bones of the carpus and metacarpus; the nails also were not widened. Among the bones of the face, the malar bones and lower jaw were specially hypertrophied; no exostoses. The cranial bones were very much thickened, which was evident at the orbital fossæ. The muscles were diminished in size, especially where the bones were hypertrophied. Veins abnormally developed, with varices. Since the commencement of the illness the sight had failed and the hearing become considerably diminished.

[1] Binet, 'Bulletin de la Société Clinique,' 1882, p. 8.
[2] Richardière-Rogier case, Thesis of Rogier, Paris, 1884.

No loss of memory or intelligence. Sensation and tendon reflexes normal. Urine healthy.

The principal characters described by Paget are found in all these cases, and are briefly the following:—Considerable increase in the size and curve of the long bones, and of the bones of the trunk and head, producing a characteristic appearance. The femurs and tibias are curved considerably forwards; the legs are abducted; and the neck and trunk are fixed in a condition of forward flexure, as in the similar curve of rickets. The respiration is impeded and almost entirely diaphragmatic in character, on account of the hypertrophy and ankylosis of the ribs. Frequently, both at the commencement and during the disease, there have been severe pains analogous to lightning pains. Its duration may be very long, and it may last ten to fifteen years without marked interference with the general health. Syphilis has no connection with the cause of this affection, which is perhaps not without some relation to cancer and especially to gout.

Let us now return more directly to our subject, and examine the relations which exist between the malady which we have described and Paget's disease. In the former as in the latter we find an increase in the size of the head and the bones of the limbs, increased development of the venous system, and muscular weakness. These analogies are, we admit, of considerable importance, and speak strongly in favour of the identity of the two diseases. At one of the meetings of the Clinical Society (July, 1885), when showing our two patients, we expressed the opinion, not without some reserve, that they were abnormal cases of Paget's disease. But on subsequent comparison we have been obliged to change our opinion, and recognise that they belong to a very distinct malady. Let us examine what the differences are. If the patient is affected by osteitis deformans, it is especially the bones of the cranium, which by their overgrowth produce increase in the size of the head. If, as occurs occasionally, the bones of the face are also attacked, it is only to a slight degree, and as one may say secondarily. In our patients, on the contrary, the bones of the face are those which specially undergo hypertrophy, those of the cranium having a less marked development. In the first disease the face has a triangular aspect with the base above, in the latter it is that of a lengthened ellipse; while finally in myxœdema it is circular (" in full moon," as Sir William Gull has aptly termed it).

As regards the locality of the hypertrophy of the limbs, it is far from being alike in the two diseases. We have seen that the special feature presented by our patients is an enormous hypertrophy of the hands and feet, coming on most often without noticeable change in the long bones, or at any rate preceding it when the latter exists; resulting in a marked contrast between these parts and the slenderness of the limb itself. In Paget's disease, however, on the contrary, it is the long bones which are chiefly affected; the hypertrophy hardly ever attacking the bones of the hands and feet, or only to a very slight extent. As to the other bones of the skeleton (clavicles, ribs, pelvis, &c.), it has been seen that they may be attacked in either affection; but we believe they suffer more in osteitis deformans than in the other morbid process. Another feature of great importance in the former affection is the tendency to deformity of the bones (whence its name), which is not observed in the latter disease. In this disease, in fact, these curves of the tibias and femurs outwards and forwards, and more rarely of the humeri and bones of the forearm, are never seen. In a certain number of Sir James Paget's cases a shortening of the height is noticed, following the above-mentioned deformities; but in our patients, on the contrary, there is more often increase in the height.

Moreover the commencement of the disease is essentially different in the two diseases: for instance, osteitis deformans does not show itself till after the age of forty, whereas the malady we have described comes on from fifteen to thirty-five years; the first is not a family disease, the second may be seen in more than one member of the same family.[1]

Finally, the invasion of different parts is not more allied. In our patients the disease begins symmetrically, the two hands or two feet being attacked at once; in osteitis deformans, on the contrary, the invasion is effected in a much

[1] Friedreich.

more disassociated manner, it being a tibia or a femur which is first attacked, the corresponding bone on the opposite side being affected subsequently, and the bones primarily involved remaining throughout the disease more hypertrophied than those of the other side.

Such are our reasons for considering that these diseases belong to two distinct groups; although there are between them several points of resemblance. We may add that Sir James Paget himself considered the case recorded by Saucerotte as a very doubtful example of osteitis deformans, and also that he expressed the opinion that our two diseases were not identical. We wish to acknowledge the kindness with which he has placed his valuable writings at our disposal. Such being the case, what should be considered the nature of this malady? The answer is difficult with so few data on which at present we can rely. We may, however, at once consider the idea of a local affection eliminated, with such disseminated lesions before us; and it may then be said to rest on three hypotheses.

A. Is it a general malady possibly more or less allied to rheumatism? The changes can perhaps be compared to a certain extent with the osseous lesions of rheumatism, noted by Adams and Fereol. Against this theory is the slightness of the pains in the limbs, the integrity of the joints and serous membranes in general, the almost stationary condition after a certain time, and the absence of exacerbations or of any acute period.

B. The manifestations of the disease may be directly dependent on lesions of the nervous system, and specially of the main sympathetic. In favour of this view are the trophic nature of the lesions, and the results of the autopsy by Dr. Henrot. It is well, however, to note in this case that though there was considerable hypertrophy of the sympathetic, nothing proves that it was the immediate cause of hypertrophy of the limbs. In fact, both may have been developed simultaneously without standing in the relation of cause and effect.

C. It may be a defect of development, attacking certain parts of the anatomical system, commencing during adolescence or early adult life, possibly a family disease; and having certain pathological analogies with another malady[1] which seems also a defect of development, *i. e.* progressive primitive myopathy.

Whether this is the case or not, and much as we incline towards this view, we cannot insist too much that they are pure hypotheses, and that on the whole we have as yet no certain data as to the nature of this malady. The latter, however, appears to us to present all the features of a true morbid entity, and thus merits a place apart from all connection with the group as yet so little known of more or less general hyperostoses.

Conclusions.

1st. There exists a disease specially characterised by an hypertrophy of the hands, feet, and face, which we propose to call Acromegaly ; that is to say, hypertrophy of the hands and feet (not because actually the hands and feet are alone affected during the course of the malady, but because their increase is an initial lesion, and constitutes the most characteristic feature of the affection).

2nd. Acromegaly is quite distinct from myxœdema, and from Paget's disease (osteitis deformans), as well as from leontiasis ossea of Virchow.[2]

[1] Very different clinically, as it is the muscles and not the bones which it attacks.

[2] Republished in the 'Revue de Médecine,' September, 1885.

A THESIS

ON

ACROMEGALY

(MARIE'S MALADY).

BY

SOUZA-LEITE, J.D.,

DOCTOR OF MEDICINE OF THE FACULTY OF PARIS AND BAHIA; MEMBER OF
THE ANTHROPOLOGICAL SOCIETY; ASSOCIATE OF THE MEDICO-
PSYCHOLOGICAL SOCIETY.

A Thesis published in 1890.

TRANSLATED BY
PROCTER S. HUTCHINSON, M.R.C.S.,
ASSISTANT SURGEON TO THE HOSPITAL FOR DISEASES OF THE THROAT.

ACROMEGALY
(MARIE'S MALADY).

CHAPTER I.

INTRODUCTION.

The disease which is the subject of this work has, so to speak, no history; or at least it may be simply summed up in the words of our title, "Marie's Malady." It was as recently as 1885 that Dr. P. Marie named and described for the first time this peculiar disease Acromegaly, which he had studied in two women patients in the practice of Professor Charcot, at the Salpêtrière Hospital. Since then this author has published other cases of this disease, which correspond exactly to his first description, and to which he has had little further to add.

The two cases described by Marie had not been the first observed, as pointed out by that author in his original work, in which he has collected all those he has been able to find in medical literature. Among these cases the oldest recorded is certainly that of Saucerotte-Noël (1772). More recently come those of Brigidi, Chalk, Verga, Henrot, &c., and that which is the subject of an important work by Drs. Fritsche and Klebs. All these cases have, however, been confounded under different names, such as general osseous hypertrophy, pachydermatous cachexia, gigantism, exophthalmic cachexia, &c.

To Dr. P. Marie is due the merit of recognising it as a distinct malady, and of describing it in such a manner that it can be identified by everyone.

A number of memoirs and contributions have since appeared on the subject; especially may be mentioned those of Erb, Virchow, Wilks, Hadden, Verstraeten, Adler, Minkowski, as among the authors who have contributed to the study of acromegaly. A review by M. G. Guinon may also be cited for the precision and clearness with which it treats of the disease.

Before studying in detail the different symptoms and modifications of acromegaly, it may, perhaps, be as well to give a brief summary of its principal features. Taking these from their symptomatic aspect, in the order in which they first attract attention, we notice, first, enlarged hands, often very distinctly widened and thickened—massive, in short. The soft structures on their palmar surface project like pads, the fingers throughout their length present similar increase, and, curiously, neither the hand, properly speaking, nor the fingers are notably lengthened. The hands are like "battledores," and the fingers may be compared to "sausages." The other segments of the upper limbs do not share in the striking changes of which we have just spoken. If, as occurs in some of the subjects of acromegaly, the limbs are more or less enlarged, this increase in size is never so considerable as that of the hands. As regards the feet and lower limbs, the same remarks apply; the dimensions of these parts, however, often show less increase than would be in proportion to those of the upper limbs.

Secondly, as regards the head, the face is altered in an extraordinary manner. It is lengthened from above downwards. The upper jaw is thickened and markedly prognathous, so that the face takes the shape of an elongated oval, and shows a tendency to become slightly concave. If to this is added the thick lips (principally the lower one, which is overhanging), showing between them the teeth and the point of the tongue, the latter more or less thickened and enlarged; a big flat nose, having thickened bridge and alæ (which are also more rigid than normal); projecting orbital ridges, surrounding thickened lids and prominent eyeballs, the latter appearing too small; a relatively low forehead, sometimes a little retreating; finally, a marked pigmentation of the skin of the face, specially of that of the lids,

—if all these details are considered, it will be seen that there is a distinct diagnostic type of face in acromegaly.

Thirdly, to these considerable changes in the extremities may be added the not less interesting modification of the trunk. This also becomes very large, and presents, when looked at from the side, (a) a very marked cervico-dorsal curvature, (b) a slight lumbar lordosis. In front is a more or less marked projection of the ends of the ribs and of the sternum, which distinctly depresses the level of the abdominal wall. Resulting from this a double hump is formed, which will be described later. Add to the above description amenorrhœa, a thick neck, alteration in the tone of the voice, and other less manifest objective symptoms, and we have an almost sufficient basis for diagnosis.

Fourthly, the long list of subjective symptoms—the headache with its varying character and intensity; the more or less severe pains in other parts of the body; the excessive appetite and thirst; the remarkable changes in the vision—all confirm the diagnosis when, after examining the extremities, any doubt exists.

The progress of all these symptoms, and their duration to the termination of the malady, though varying with the individual case, are usually long, ten to twenty years or even more. It is one of the most chronic maladies. The cause of acromegaly is not yet known.

CHAPTER II.

ETIOLOGY.

IN spite of recent researches, the knowledge of the causes which produce Marie's disease is still indefinite. This is specially due to the fact that acromegaly commences insidiously, so that the subjects of it overlook the onset of their malady. Often, indeed, they are not aware of the increase in the size of their hands and face till their attention is called to it by their medical adviser. We may, however, review a certain number of causes, the importance of many of which is very slight. The influence of Heredity has not been noticed to play any part in acromegaly; it has not been recorded in any known case.

As regards Race, further information is wanting before arriving at definite conclusions, but it seems to exert no influence. Cases of acromegaly have now been published by doctors in nearly all countries. Both sexes may be attacked; the malady is nearly as frequent in men as in women (16 cases concern men, 22 women). Age, on the contrary, is of special importance. It is about adolescence that the malady generally commences; in fact, in the majority of cases of which the onset has been definitely noted, it is between 19 and 26 that the first symptoms have made their appearance. In Freund's case (Case XIV) the disease showed itself at puberty. On the other hand, in Professor Erb's case (Case XX) the acromegaly did not appear till the age of 48, at the time of the menopause. These two cases are, however, exceptional; and the later years of adolescence, and the first part of adult age, would seem to be the time when patients are most frequently attacked by the disease.

Depressing mental emotions, such as grief, lassitude, &c., have been recorded in Dr. Péchadre's case (No. IX), where the first symptoms of acromegaly came on between 35 and 36 years, coincident with the attacks of depression. General chill, and particularly of the extremities, has been noted (Cases VI, XII, V); but this is a cause common to so many maladies that it is difficult to assign its true importance. The same may be said of traumatism, noted in Cases VI and XXVI.

Rheumatism and gout, the two chief constitutional maladies to which the most important manifestations of the arthritic diathesis are due, and themselves the causes of several diseases, are mentioned in Cases I, IV, XXII, and XXIII. Mr. Godlee's patient is the offspring of a family in which are found both gout and rheumatism. The cases, however, where these two maladies do not occur are very numerous, so much so that it may be safely asserted that these two morbid processes do not exert a very great influence on acromegaly. In all the cases the facts show that there are certain relations between the diseases called arthritic and Marie's malady. Syphilis is noted in Cases IV, XXI, and VI. The subject of Case IV, M. C—, contracted this malady at the age of 23, eight years before the appearance of the first symptoms of acromegaly. As regards M. C— it may be noted that between 17 and 18 his height increased to a considerable extent, and that ever since then he has not ceased growing, slightly it is true, yet definitely (Marie). This history shows distinctly that syphilis did not in this instance hasten the onset of acromegaly. Ghirlenzoni (Case XXXI) contracted syphilis between 35 and 40; and when he arrived at Florence in 1850 (aged 49), after his theatrical tour, he was already the subject of deformities. The causal influence of syphilis, therefore, is far from clear in this particular case. Moreover, Brigidi states of this patient that while he was in good health he was addicted to alcohol, so that alcoholism might be included as a possible cause. As regards Héron (Case VI), she had had multiple abscesses at the age of 30, and former syphilis. This patient appeared also to have had some years before an attack of variola and articular rheumatism. In Case XXII scarlatina, which gave rise to unusual symptoms, is noted in the antecedents. However, it must not be forgotten that the menses had ceased in this patient some months before the appearance of scarlatina. The onset of acromegaly was in this case accompanied by certain symptoms, which it may be well to revert to. At the commencement of 1881 there was the sensation of pricking and pain in the hands; at the end of this year final cessation of menstruation. During the first months of 1882, ordinary scarlatina; and during the remainder of the same year onset of the first symptoms of acromegaly in the hands. The etiological relation of scarlatina is thus very doubtful.

Intermittent fever and bronchorrhœa are noted in Case XXXIV; the patient had been the subject of attacks of ague with hæmorrhage from the lung, but their relation to the cause of acromegaly is doubtful. All the causes which have been enumerated may be considered rather as occasional determining factors in hastening the onset of acromegaly rather than as true predisponents.

M. P. Marie, notwithstanding his competency, states that he has formed no definite conclusions as to the etiology of this affection. It would appear to commence insidiously at adolescence or adult age. It is neither congenital nor hereditary. M. P. Marie considers, moreover, that those cases which have been described as congenital acromegaly should be excluded from the category of the true disease.

CHAPTER III.

SYMPTOMS.

In acromegaly, what first attracts attention in inspecting the patient is the great size of the hands, feet, and head; and also the deviations of the spine, principally its upper half. In order to see clearly the different features of the disease it may be useful to arrange them in the following manner: 1st, objective symptoms; 2nd, subjective symptoms; 3rd, general symptoms. Each of these primary groups may be again divided in turn into (*a*) symptoms which are constant and fundamental, (*b*) symptoms which are inconstant, accessory, or secondary.

Constant Objective Symptoms.

We may commence with the hands, for it is in them that the onset of acromegaly ordinarily appears. When the hands, including the fingers, are examined in the subjects of acromegaly, attention is at once attracted to their increased size, which makes them contrast strongly with the almost normal shape of the other parts of the upper limbs, the arm and forearm. The excessive development of the hands includes all the tissues composing them, bones, muscles, connective tissue, and skin. A general hypertrophy is the result, which involves the width and thickness of the fingers and entire hand. The length of the hand measured from the wrist to the end of the middle finger is about the same as in a normal hand; whereas their width and thickness are enormous. The terms "battledore hands" (Marie), and "spade-like," of English authors, express their appearance.

The consistence of the hands thus hypertrophied is firm,

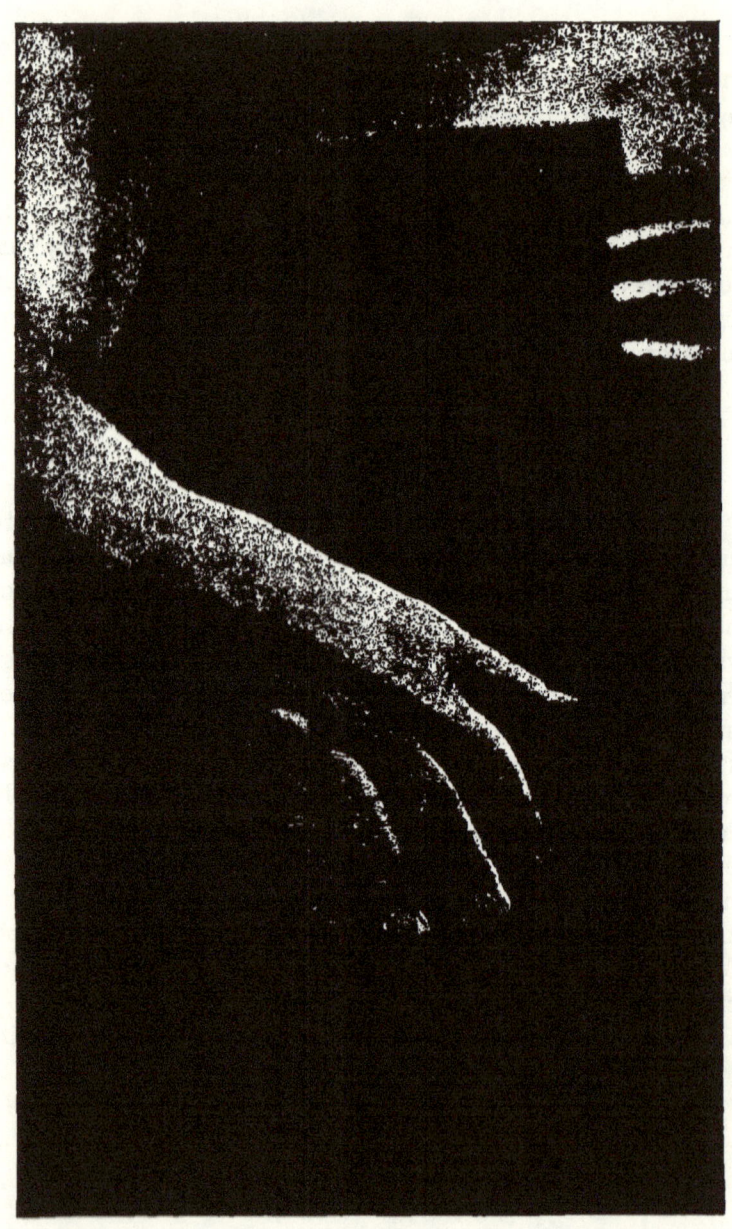

Fig. 4.—Hand of patient, the subject of Case III.

and gives none of the pitting on pressure characteristic of œdema. On examining the hands closer the interphalangeal folds are found to be deeper than normal. The thenar and hypothenar eminences are also abnormally prominent, and increased in size; thus making deep tortuous folds in the palm

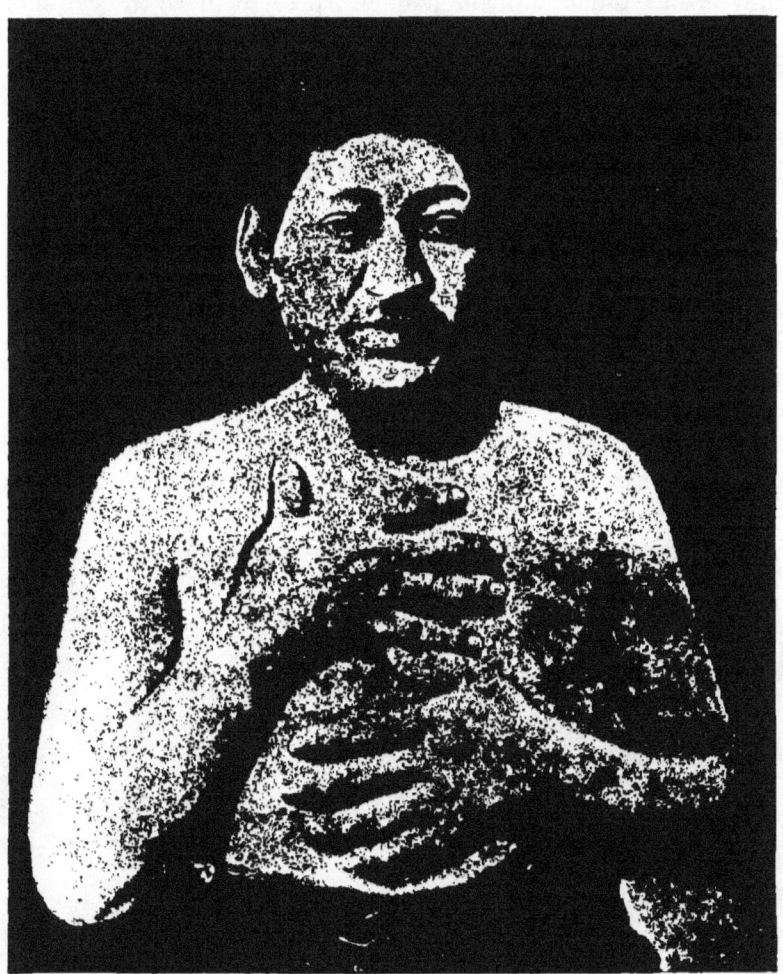

Fig. 5.—Hands of patient, the subject of Dr. Roth's case.

of the hand, proportionate in depth to the development of these eminences. The fingers as well as the palm of the hand assume enormous proportions. They preserve their natural straightness, and are as large at their tips as at their base.

Their shape is not quite cylindrical, being flattened from before backwards. There is no deformity at the phalangeal joints. The cutaneous folds opposite the joints are deepened, making projecting ridges between them. The nails are small in comparison with the size of the corresponding fingers; they are flattened and widened, and present in all the patients a striation which is nearly always longitudinal. This striation was transverse in the subject of Professor Verstraeten's case, who believed that this change commenced in the outer fingers of the hand. Nodosities opposite the first phalangeal joint have been noticed in some of the patients, which have the appearance of the nodosities which occur with dilated stomach, observed by Professor Bouchard. Before finally leaving the hands, we may note that these segments of the upper limb, as well as the feet, were increased in length in L— (Case XII) and in H— (Case XIX); as regards L—, we must not forget that his acromegaly was in process of development.

We have observed above that this hypertrophy of the hand does not involve, as a rule, the arm or forearm; and that it

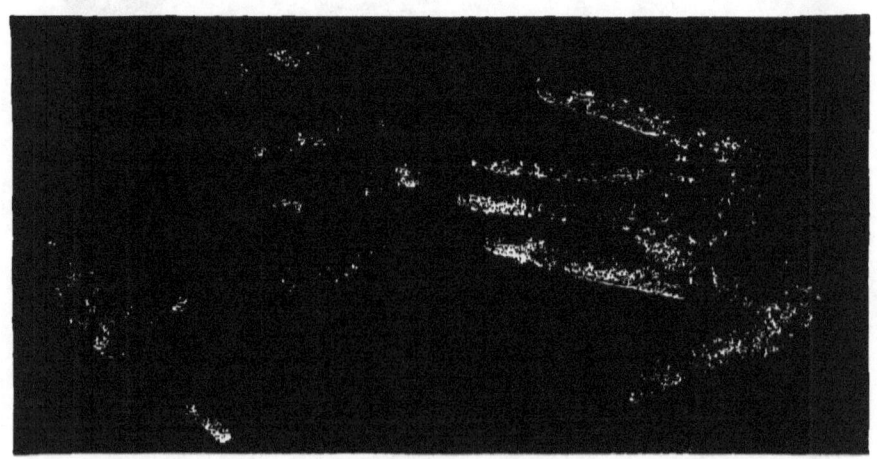

Fig. 6.—Hand of patient, the subject of Case I, compared to that of a well-developed man.

stops at the wrist in the great majority of cases. In a few, however, these parts may be increased in size (Cases I, III,

IX, XX, XXVI). In certain individuals (Virchow, Marie) the muscular power was remarkable, and the arm and forearm were of large size; their dimensions, however, never corresponded to the hands. The feet are affected in the same way as the hands; that is to say, they are enlarged, thickened, and flattened, at the same time preserving their normal length. The same ridges and furrows are also present, the one between the sole of the foot and the digits being specially marked.

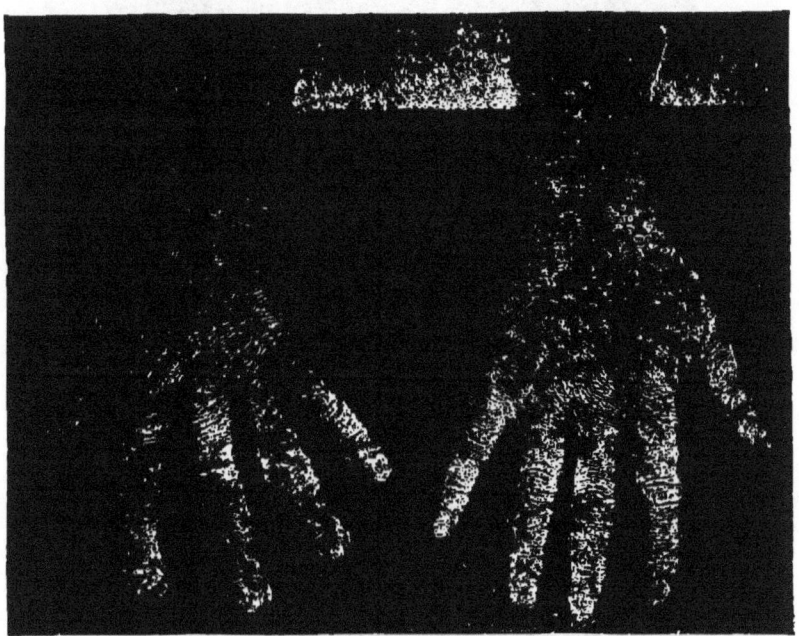

FIG. 7.—Hands of patient, the subject of Case I, compared to that of a well-developed man.

In the feet, as in the hands, the hypertrophy does not extend above the ankle, and the legs may present an appearance of atrophy in comparison with them. The other observations made regarding the hands apply also to the feet.

The head.—The head presents, like the hands and feet, a very distinct increase in size. The face of the subjects of acromegaly is, in fact, so marked that they can be recognised at first glance. Examining the head closely, however, it is found that all parts are not equally affected. The cranium is but little attacked by the disease, the chief changes taking place in the face. The face is enlarged, especially in the vertical diameter, being elongated and oval. The forehead is low in nearly all cases, and is supported on enormously thickened orbits, which cause it to be partly obscured, also making its shape appear altered. The borders of the orbits are also increased in consequence of dilatation of the frontal sinuses surrounding the orbital cavity. The con-

Fig. 8.—The woman who is the subject of Case V. The oval of the face is distinctly increased.

junctiva, fat, and muscles of the eye may be somewhat modified (Cases I, XXXII, XXXIII); and exophthalmos may sometimes be more or less pronounced (Cases II, VI, XXVII). Sometimes the eyes are small, expressionless, and out of proportion to the size of the orbits (Cases XXXI, XXXII). The anterior part of the temporal region, just behind the orbits, may be depressed so as to resemble this region in the cow (Marie). The lids are long, thicker than normal, and their colour somewhat brownish; in some cases they incompletely cover the eyes, particularly if there is exophthalmos. This thickening of the lids involves their whole length, the tarsal cartilages being sometimes definitely hypertrophied. The nose is one of the most hypertrophied parts of the face; it may become most unusually large. It forms a marked projection in the middle of the already enlarged face, and gives a strange appearance to the physiognomy. It is enlarged in all its dimensions. The alæ are thickened and large, especially at their lower part; the septum may be double its normal thickness.

The cheek-bones are prominent, and form, with the enlarged superciliary ridges, a projection all round the orbits. All these prominences, which have been mentioned in regard to the face, are due to dilatation of the various sinuses of the bones.

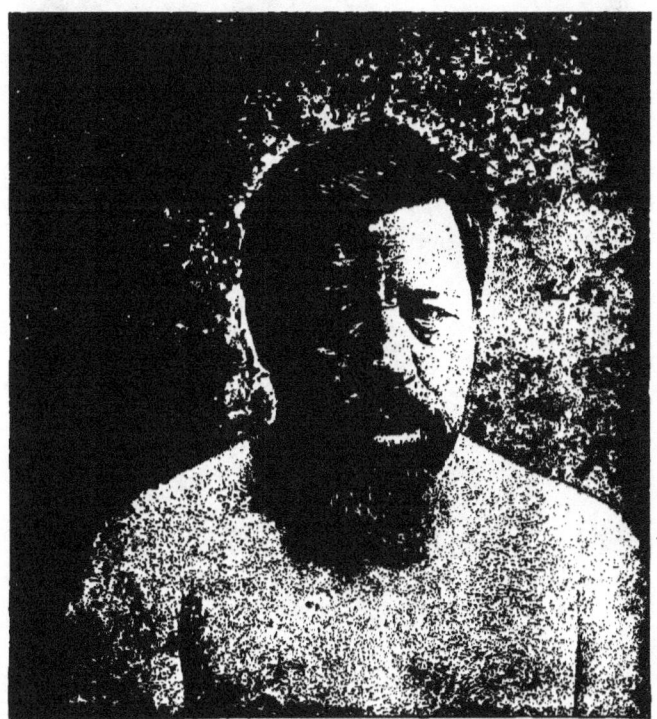

FIG. 9.—The patient who is the subject of Case III. The size of the lower lip is very marked.

The lips are specially to be noted. They are swollen, but not to an equal extent. The upper lip is less hypertrophied than the lower, and appears small by comparison with the latter, which is prominent, turned downwards, and overhanging (Cases III, XXII, XXXI, XXXII). With this hypertrophy of the lips there is frequently present an open mouth, which allows the hypertrophied tongue to be seen. The latter is large, increased in width and thickness, but preserving its normal shape. The tongue may be so large as

Fig. 10.—The patient who is the subject of Case IV; his height was 1 m. 80. The length of chin is very marked.

to project from the mouth. Its size interferes with its movement, and makes the pronunciation of words difficult. On examining the inside of the mouth it is found that the palatine arch and soft palate are enlarged, especially from before backwards; the tonsils and pillars of the fauces are also increased in size, making the voice guttural and a little metallic, as in the patients of MM. Hadden and Ballance. The uvula is sometimes elongated and enlarged, exciting cough. The chin especially shares in the general increase of the face; it is large and massive, forming a distinct projection directed downwards and forwards, which helps to give the face its strange appearance. All the soft structures of the face are also thickened, and the sinuses of the bones dilated.

The upper jaw is but little modified as compared to the lower. This inequality in development explains the prognathous type of jaw, which is sometimes considerable in acromegaly (Cases XXXI, XXXII, XXXVII, XXXVIII, VI, I).

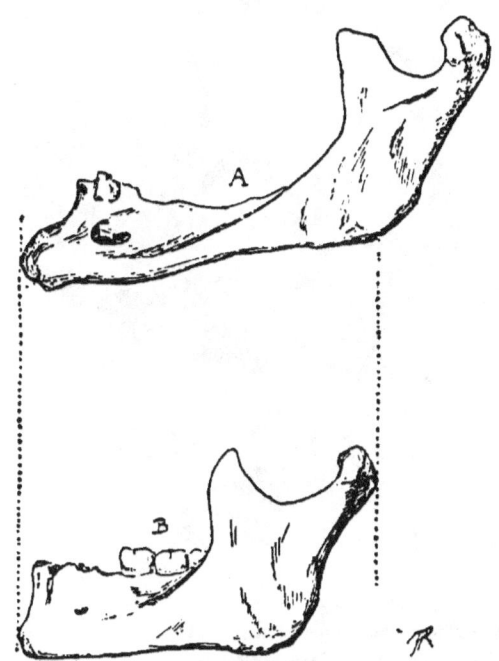

Fig. 11.—A is the jaw of the patient who is the subject of Case VI; B, a normal jaw (design by F. Richer).

The teeth, as a result of this, are separated more or less one from another, and the upper dental arch tends to fit inside the lower, the latter projecting some distance in front of the former. The alveolar borders, particularly the lower, may be found more or less atrophied (Cases V, VI, XX, &c.), in consequence of the falling out of the teeth (atrophy (?) and caries). In no case of acromegaly have we found the hypertrophy of the teeth recorded by Dr. Henrot.

An increase in the size of the ears, varying in different cases, may also be observed. Their cartilages may become thicker and firmer; in the same manner, also, those of the nose and larynx.

The face is found, from all these different alterations of its parts, to have assumed the form of a long oval, as has been well described by Marie.

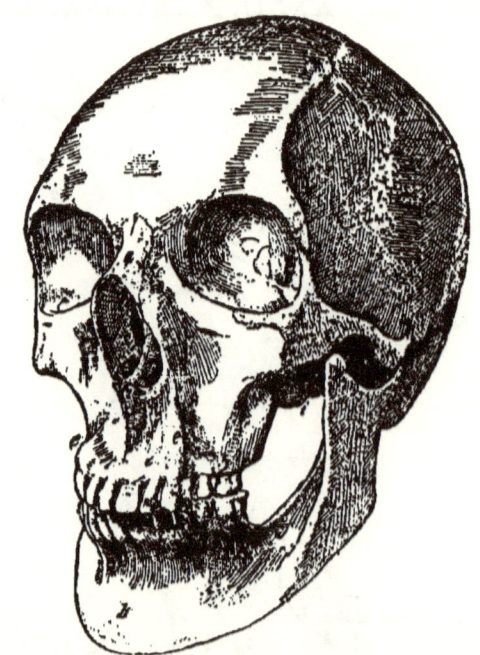

FIG. 12.—The woman observed by Dr. Verga; lower jaw large; face lengthened.

The cranium is not perfectly normal; it becomes somewhat increased in its antero-posterior diameter. The modifications in the cranium are chiefly owing to dilatation of the frontal sinuses. Dr. P. Marie has observed thickening of the bony crests in certain cases. Opposite the cranial sutures the occipital protuberance may become thickened, and form a crest. The same also may be observed over the mastoid processes (Schültze's case), but these alterations are very slight compared with those of the face.

Before leaving the condition of the extremities in acromegaly, we may state that the onset of the deformities which we have described, instead of first attacking the hands as is generally the case, may involve the head first, and more especially the face (Cases IV, VI, XII, XXIII).

We may now mention the changes in the thorax, and will commence with the spine. The spine becomes deviated to a variable extent. This deviation has been recorded in all the cases, in those at least in which acromegaly had become fully established in the patients. We consider it should rank as of equal importance with the deformities of the extremities, and be classed among the fundamental symptoms. The deviations of the spine consist of curvature in the cervico-dorsal region. The back is rounded, and the patient holds the head straight with difficulty. The latter becomes continually bent forwards, and appears to recede into the shoulders, particularly if one of these is higher than the other. A lateral curvature is frequently also present, with its concavity either to the right or the left. A certain degree of lordosis may also be found, which compensates somewhat for the first of these deformities, and is usually situated in the lower part of the dorsal and lumbar regions. The spines of the vertebræ, so far as they have been examined in the living subject, are hypertrophied. The thorax is enlarged in nearly all the bones which enter into its formation. It projects in front, its antero-posterior diameter being increased to a certain extent, apparently at the expense of its transverse diameter. The thorax, as a result of this, is flattened from side to side; the sternal portion is directed downwards and forwards. The lower part of the thorax also projects in a peculiar manner, which causes a marked prominence when the patient inspires deeply. The sternum is one of the bones of the thorax most affected, being thickened and widened, and also a little longer than normal; there are crests at the junction of its various divisions. The xiphoid cartilage is less flexible than normal. The clavicles become larger, particularly at their extremities. The ribs are thick and wide, including their angles. The costal cartilages are

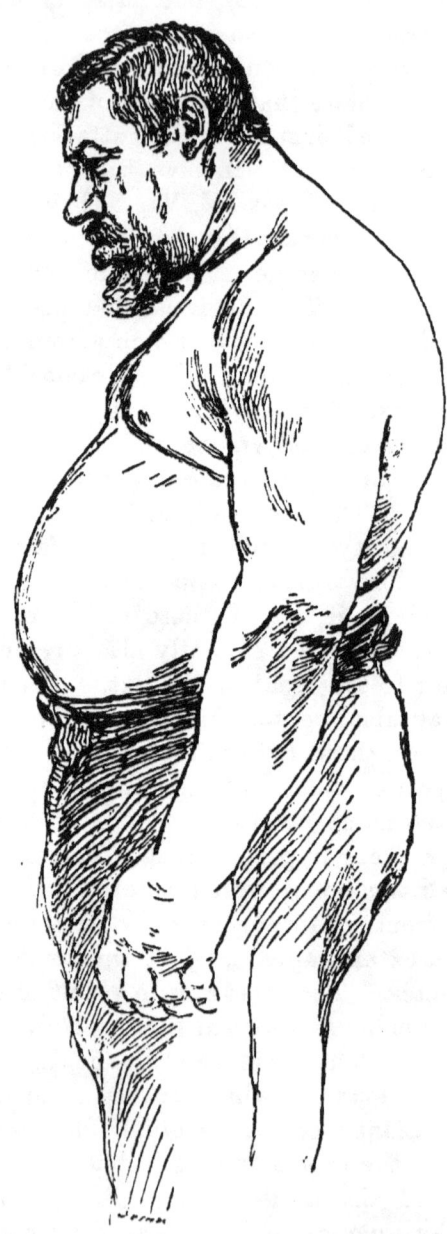

Fig. 13.—The patient who is the subject of Case III.

enlarged, and somewhat irregular to the touch. They are ossified to a variable extent, and present in some of the subjects of acromegaly a beaded condition like that of rickets. The xiphoid cartilage is ossified in some cases.

The lower ribs become much inclined downwards and forwards, particularly when the xiphoid cartilage is longer than normal. When the patient is sitting with the head bent forward, the ribs may then touch the iliac crests. The scapulas are often thickened, particularly their lower borders, spines, acromion, and coracoid processes.

During the movements of respiration the upper part of the thorax moves but little; and these movements are mainly accomplished by the lower part of the thorax and abdomen.

All these changes in the thorax above mentioned are produced, as elsewhere, in a slow and progressive manner. They may appear at the same time, or, what is more frequent, a little later, than the changes in the extremities.

From this description of these thoracic lesions it is seen that two humps are formed, one in front and one behind, like that usually represented on "Punch" (Marie). Such are the characteristic and specific deformities of acromegaly, which are constant and form the basis of the disease. There exist, however, others more variable in frequency and intensity, and involving several different organs.

Secondary and Inconstant Objective Symptoms.

The neck is often thick, comparatively short, and, together with the cervico-dorsal curvature and the raising of the shoulders, combines to make the head appear very low. This enlargement of the neck is not due to any increase in the thyroid gland, but to true hypertrophy of the tissues of the neck. The thyroid body, in fact, is diminished in size in the majority of patients, though a total disappearance has not been observed. In some it remains normal, and in a few hypertrophy exists. The larynx may become thickened in a marked degree, not only in men but also in women. In the latter the pomum Adami may be as much developed as in man. The cricoid and arytænoid cartilages are thicker and more resistent than normal, sometimes also partially ossified; while the speech becomes slow, thick, and guttural, and the voice modified in a peculiar manner. The latter is almost always deeper, stronger, and harsher; sometimes sounding very disagreeable. According to Marie, this change in the voice is due to the resonance given it by the dilated sinuses of the face. It is probable, also, that the increased size of the interior of the larynx and its ventricles, and also the thickening of the vocal cords, assist in its production.

The area of dulness over the upper sternal region, described by Erb, and mentioned also by Schültze and Verstraeten, has been attributed by these authors to persistence of the thymus gland. The foundation for this view of its nature seems, however, doubtful. Marie and the author (Souza-Leite) have observed slight dulness at the side of the sternum in only one case.

The mammary glands may be small, soft, and atrophied. The nipple, on the contrary, is large, and is surrounded by thick hairs. The abdomen is more or less large and pendulous. The skin over this region may be thickened and crossed transversely with furrows, which obscure the umbilicus.

The pelvis may present enlargement in all its dimensions. The iliac bones are thickened, and the crest, iliac spines, ischia, and horizontal rami of the pubes are all more easily detected than normal; the mons Veneris is more prominent than usual, the buttocks are lax, the sacral region is a little flatter and less curved than normal, tending to assume a vertical direction; also the sacral crest is more noticeable to the touch. The sexual organs are sometimes the seat of very important changes. The labia majora, nymphæ, and other parts of the vulva may be thickened; the clitoris in particular may be very large. The vagina in these cases is long, and remarkably increased in size. The uterus is small, raised, and shows some signs of senile changes. Its walls are flaccid and somewhat thin. In Freund's case, the ovaries were flattened, firm, and irregular to the touch.

In the male, alteration in the size of the penis may be

noticed, which becomes thicker and longer. The scrotum and testes may show hypertrophy, or, on the contrary, may be atrophied. The morbid process involves the urethra, which becomes tortuous, and may be as large as the little finger, forming a urethral "cord." The muscles of the limbs have been found sometimes atrophied and sometimes hypertrophied, according to the duration of the malady. Thus the muscles may be normal or over-developed, with a corresponding increase in muscular power. At other times they are diminished in size, flabby, and less firm than normal; in which case the patients are feeble and easily tired. The latter is the more common condition. In Erb's case the electric excitability of the muscles was increased. The interossei muscles seem atrophied in a few cases (Cases IV, VI, XXVIII). The joints may be a little enlarged, especially those of the knee, wrist, and ankle; but they are never deformed. The patellas also may be thickened, so that the knees project at the side or in front (Cases I, V, VI, XXVIII).

Cracklings and friction-sounds may take place spontaneously, and during passive movements of the joints, as in Case V. They give almost the sensation of a bag of nuts. The joints, however, present none of the exudations or other features characteristic of rheumatoid arthritis. The long bones of the limbs are in general little altered in their shape or normal direction; and when any change is noted it is very slight, thus essentially differing from those of osteitis deformans of Paget. The limbs are occasionally enlarged, but this is due to a true hypertrophy of all the structures forming them.

The movements and muscular power of the limbs are often diminished at a late period of acromegaly; the patient then becomes feeble and incapable of effort, and takes to his bed. The tendon-reflexes, in particular the patellar reflex, are present in most of the patients; they may, however, be diminished or even absent, but have not been found increased.

With regard to the vascular system, important changes have been noted. The heart is increased in size. This hypertrophy was more marked in Freund's case than in that

Fig. 14.—The wrestler recorded by Dr. Virchow. The muscles present a considerable development.

of Erb, in whose patient a systolic bruit was heard at the apex. A bruit was also noticed in Klebs' patient. Those arteries which are accessible to the touch present a certain degree of rigidity in their walls, like vessels in the first stage of atheroma. Among these are the temporals, brachials, and radials. It is in the veins, however, that alterations are most commonly found; varicose dilatations of veins are, comparatively to the above-mentioned changes, more frequently met with in acromegaly. They involve those of the hands, the legs, and the rectum. Hæmorrhoids become much developed, and cause more or less troublesome and persistent hæmorrhages. Some patients suffer from slight hæmorrhages from the trachea or bronchi. The lymphatic vessels and glands are sometimes involved. This was specially the case in Henrot's patient, in whom the submaxillary and parotid glands were also hypertrophied.

Perspiration is much increased in the majority of cases of acromegaly. The least exercise or manual labour, a short walk, or the warmth of bed, provokes abundant sweating. In women these sweats become more profuse as soon as the suppression of the menses has become permanently established. The sweat may have a disagreeable odour.

The urine may in certain cases be very abundant, amounting to a marked polyuria. This polyuria may be present by itself, or combined with glycosuria (as in two cases of Marie's and one case of Strümpnell, Cases III, V, and XIV). This presents a special interest, as it seems to be a corollary of the disturbances of nutrition, which are experienced in such different and severe forms by the subjects of acromegaly. In both Marie's patients this glycosuria was clearly connected with the food, for it disappeared on subjecting the patient to a strict diabetic diet. The influence of each repast was distinctly noticeable on the quantity of sugar in the urine. It was reduced in the night in a very perceptible manner. The presence of albumen has also been observed, the amount being very variable (nephritis ?). One particularly interesting fact is the presence in the urine of peptones. This was first noticed by Professor Bouchard; since then it has been found in two other cases by Souza-Leite and Grenvillet. Affections of the cutaneous sensibility are rare.

Anæsthesia and analgesia were detected by Strümpnell, Erb, and Henrot (Cases XVI, XX, XXXII), but neither was marked. Verstraeten's and Erb's patients both showed a certain degree of increased susceptibility to cold. The peripheral temperature may be lower than that of the axilla (Case VIII), and the patient may occasionally complain that his feet are hotter than before the commencement of his malady.

The symmetry of the body may not be preserved, one or other side having been observed by Verstraeten to be unusually developed. This fact has not been noticed by other observers, and calls attention to the result of former observations, i. e. that the differences of this kind are very much more frequent in healthy individuals (Guinon). We have already noted that the skin of the hands and feet is of deeper colour than the rest of the limbs and trunk. The skin may also take on a yellowish-brown or olive tint. The skin becomes in certain cases a little thick, dry or greasy, and as if too large for the body (Cases XIV, IV, XXVIII, XX). It may also be raised into great thick folds, and presented in one of Marie's patients the condition of goose-skin.

A certain number of patients present on the body, especially on the upper part of the trunk, little cutaneous tumours, sometimes pedunculated. They are the size of a millet or hemp seed, or even larger. They are of a red colour, sometimes violet, and very numerous. The condition presented is that of molluscum fibrosum or pendulum. The patient of Verstraeten presented numerous flat warts round the neck and waist. At first Marie was inclined to regard molluscum as a simple coincidence of acromegaly, but having since found it in all his patients, he questions if it is not one of the true phenomena of acromegaly, due to changes in the nutrition of the skin. The hair is abundant, thick, and grows with normal rapidity. The hairs of the body and pubes are strong, probably more thickened than normal, but not more numerous. In women the little hairs round the nipple and between the breasts may be observed to be larger and longer.

Constant Subjective Symptoms.

These phenomena are not less interesting than the objective symptoms. Among them the most prominent is the cephalalgia. This is, in fact, very frequently present, and it is for that reason that Marie has insisted on its importance. The headache of acromegaly is often the first symptom of which the patient complains, and may be the only one for which a doctor is consulted. The headache may be intense and remittent, or occurring at longer or shorter intervals. It may involve the whole head, but is usually situated in one part of the cranium, and most frequently in the occiput. In one patient it involved the whole of one side of the head, and became very severe at night. It may be modified by the position of the head. Finally, authors have noted its absence in some cases.

Besides headache, patients often complain of pains of variable character and degree, which are situated in the limbs, more particularly in the bones or in the joints. These pains, which, unlike the headache, are not ordinarily persistent, may arise spontaneously, or be provoked by the shaking of a carriage. In some cases their commencement is coincident with the appearance of deformities in the extremities.

A second symptom of the malady is the changes in the menstruation. Marie has rightly drawn attention to the importance of disturbances of this function. Often it is one of the first phenomena in the development of acromegaly; when the other changes in the hands, feet, and face have not appeared, or at least are not sufficiently marked for us to be certain of their existence, the patients have been deceived into believing that it was the commencement of pregnancy. Sometimes amenorrhœa marks the onset in a definite manner (at the age of twenty-nine, Case VI; at thirty-nine, Case XXVII; at thirty-two, Case XXII; at thirty-five, Case IX). More often the final cessation of menstruation is preceded by attacks of suppression of longer or shorter duration, varying for ten or twelve months before the final arrest. It is very probable that ovulation is much disturbed, for it seems certain that the ovaries undergo, like the uterus, a premature atrophy. Women, moreover, the subjects of acromegaly, cease to have children when their malady is established. In fact, amenorrhœa is a symptom almost constantly present in acromegaly.

In the same category may be mentioned diminution in the sexual desire. The virile power is usually diminished in men, but, as will be seen in the cases narrated, to a variable and not constant degree.

As regards the organs of special sense, lesions of the eye are among the most interesting. Sight is ordinarily feeble, but varies in degree from slight amblyopia to complete loss of sight in the later stages of acromegaly, when the patient becomes helpless and confined to bed. This amaurosis is due to definite changes, to be detected with the ophthalmoscope, both in the retina and optic nerves. It varies from simple congestion of these structures to a confirmed neuro-retinitis. Certain patients complain of severe pains in the interior of the eyes, and around these organs; especially when they have been concentrated on any object for some time. A marked irregular narrowing of the field of vision was noticed in Case XXVI. The patient of Fritsche and Klebs (Case XXVIII) was hypermetropic, and had slight exophthalmia. The latter is common in acromegaly on account of changes in the floor of the orbit. Brigidi's patient (Case XXXI) had, however, small eyes with projecting superciliary ridges. As a result of autopsies on cases of acromegaly, we know exactly to what lesion these affections of the sight are to be attributed, *i. e.* hypertrophy of the pituitary body. On account of the situation of this tumour the optic nerve and commissure are more or less pressed upon. From this pressure neuritis follows, which sufficiently explains the different disturbances of sight. A temporal hemianopsia was observed in Schultze's case. In many of the cases the appetite is noticed to be very large, and less frequently perhaps the thirst is intense. The increased appetite, Marie observes (loc. cit.), is so marked in some patients (Cases I, V, VI, III, XII, &c.) that they appear to be hungry an hour or even half an hour after a repast. Their thirst may be so difficult to quench, that a patient may drink four or five pints of water at one time. In Mr. Godlee's patient the thirst was excessive and the appetite much diminished. From the fact that the digestive functions are carried on more rapidly than normal, we may explain the absence of dilated stomach in some of the excessive eaters. In all these cases we have observed the true polyuria, the possibility of diabetes, albuminuria, and other morbid disturbances of the kidney.

Inconstant Subjective Symptoms.

These are perhaps of less importance than the corresponding objective symptoms, being more uncertain and variable than the latter. The ear is affected to a varying extent. Complete deafness may be observed, or a very slight diminution of the hearing. The patients may complain of buzzing or tingling in the ears, at the same time stating that they hear worse since their malady began. The cause may, however, be an affection of the middle ear, of the Eustachian tube, or a purulent catarrh of longer or shorter duration. An alteration in the sense of smell has been noted in some of the subjects of acromegaly. This may be disturbed for delicate odours, and a slight anæmia has been recorded in other cases. Affections of the taste have also been less frequently present: they were met with in the patients of Godlee (Case XXIII) and Strumpnell (Case XVI).

Two subjects of acromegaly complained of palpitation of the heart.

Two patients also had severe intra-abdominal pains, which radiated from the lumbar regions.

General and Mental Symptoms.

Most of these have already been mentioned, so that they may now be only briefly dealt with. The subjects of acromegaly sometimes complain of general feebleness, lassitude, and inability to work; and also a strong desire to recline. Their spirits are dejected and their temper irritable, which are probably due to the progressive weakness of sight and hearing, as well as to the other deformities. This results in a condition of melancholia, and sometimes a tendency to suicide. Ghirlenzoni (Case XXXI) and Maria B. (Case XXXV) had delirium and coma before death. Wadsworth's patient (Case XXVII) was heavy and drowsy; B. (Case III) had disturbed sleep, with bad nightmares, in which he talked loudly and gesticulated. The above are some of the numerous clinical symptoms and lesions which have been observed in acromegaly.

CHAPTER IV.

DURATION, TERMINATION, AND PROGNOSIS.

BEFORE speaking of the progress of acromegaly, we may enumerate some points concerning its onset. In the 38 cases of acromegaly mentioned in the present work two require special attention. One is a patient of Freund's, aged forty-nine, in whom the onset of acromegaly coincided with the advent of puberty. The other is Erb's patient, aged fifty-nine, in whom acromegaly had been present about twelve years. The first affords an example of early, the second of late development. Of the 38 cases above mentioned, in 14 the onset of acromegaly was between twelve and thirty years of age, in 10 the malady did not show itself till after thirty. As a general rule the malady appears during adolescence or at the commencement of adult age. The period of greatest frequency Guinon has observed to be between twenty and twenty-six years. In women the initial period may be easily determined by the amenorrhœa, which is the first symptom.

Marie entirely rejects the supposition that this affection is congenital. According to him the hypertrophies of the hands, feet, or head, which have been recorded in infants, do not come in the category of this malady. This disease becomes established slowly by progressive hypertrophy of the hands, feet, and face. Certain incidents sometimes denote the onset of acromegaly. Thus the patient may be obliged to remove his ring, or have one made larger; the first having become too small for his finger. In others the hat may require changing, and frequently also it is the gloves and boots. The patient of Minkowski's, who was a musician playing the violin, was obliged to abandon his instrument and take to a flute, as his fingers became too large; he had also subsequently to abandon the latter instrument on account of thickening of his lips.

Sometimes the patients state that their friends have for a long time exclaimed at the size of their hands; at other times they complain of pain, weakness, and lassitude. Once established, the progress of the malady is very slow, but not always equally so. Exacerbations, or periods of comparative arrest, may be noticed; but the former always preponderate over the latter. Little by little the hands and feet acquire their large and almost monstrous size. The lower half of the face and the prognathous jaw become prominent, also the humps and other deformities of the chest and head, till the subjects of acromegaly become unrecognisable. The disease lasts ten, twenty, thirty, or more years. A true cachexia may supervene towards the last stage of acromegaly. The skin becomes very loose, yellow or brown; losing its elasticity, and becoming wrinkled and scaly. The muscles become soft and emaciated. The patients, with their whole figure deformed, their hands excessively large and horny, and with a truly hideous appearance, are finally confined motionless in bed. The subjects of acromegaly finally succumb to their long and inflexible malady, from syncope, and probably also from cerebral compression. Some intercurrent malady may shorten the duration of the acromegaly.

CHAPTER V.

PATHOLOGICAL ANATOMY.

CLINICALLY acromegaly is without doubt a morbid entity; the pathological anatomy also confirms fully the distinct nature of acromegaly. Up to the present there have been seven autopsies, a number which, although small, is sufficient to establish certain changes as constant.

The most specific of these lesions, one which may be considered as essential, since it has not been found absent, is the considerable increase in size of the pituitary body. This gland is changed into an hypertrophied mass, of which the size varies from that of a pigeon's egg to that of a hen's egg, or even an apple. "M. Henrot found at the base of the brain in the middle line, and occupying the site of the pituitary body, an ovoid tumour the size of a hen's egg. It was limited by the following organs:—in front the commissure of the optic nerves, which was completely flattened, and formed a large ribbon-like band; at the sides were the hollowed-out sphenoidal lobes of the brain; behind, it rested on the cerebral peduncles. On raising it carefully it was found in close connection with the base of the brain. It was in direct continuity with the tuber cinereum, which was torn on making slight traction. The infundibulum was firm and much more developed than normal."

In Brigidi's case (No. XXII) "the gland weighed fourteen grammes; it was of irregular spheroidal form, with an antero-posterior diameter of 29 mm., and transverse 38 mm., and its thickness varied from 12 to 19 mm." In another case the pituitary tumour was as large as a hazel nut, and filled the greatly dilated sella tursica, forming a hemispherical projection. The right optic nerve was a little more compressed than the left, which was greatly flattened into the form of a band a centimetre wide. The case we have just mentioned was the patient of Fritsche and Klebs, in whom the infundibulum and pituitary body were hypertrophied.

In the autopsy of Héron (Case VI) made by Marie, the pituitary tumour measured in its transverse diameter 38 millimetres, and in the antero-posterior 32 millimetres. The

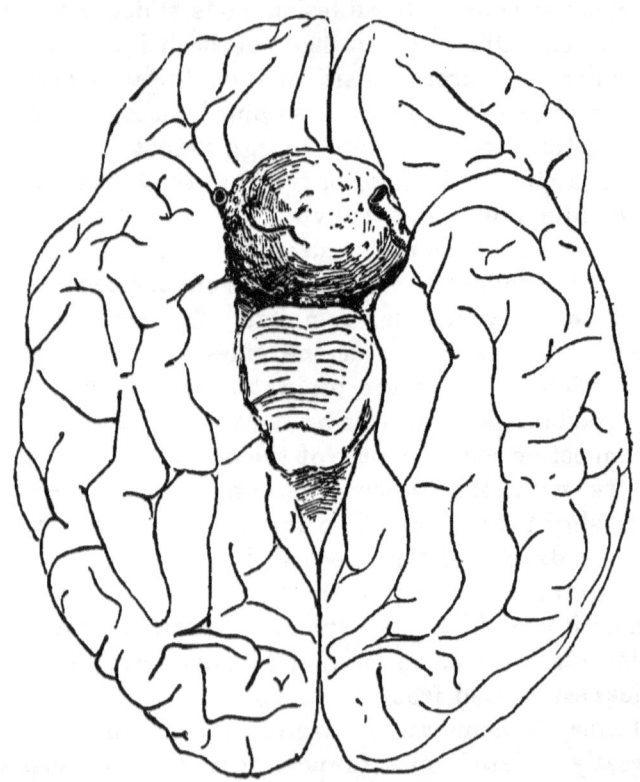

Fig. 15.—Hypertrophy of the pituitary body of the patient's brain who is the subject of Case VI (autopsy by P. Marie). (Design by Dr. Collier.)

optic nerves, besides being compressed opposite the optic bands and commissure, are also pressed upon where they enter the optic foramens. From this results the neuro-retinitis, which finally ends in complete amaurosis, and also the intra-ocular pains. From this also arises, though less manifestly, certain auditory and olfactory troubles. We believe that by this, moreover, are explained the dreams and restless nights of some patients.

Microscopically everything appears to indicate that it is not a neoplastic process in the pituitary body, but a simple

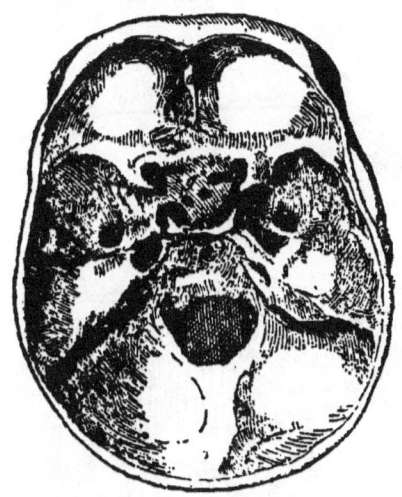

FIG. 16.—Base of cranium of Dr. Verga's patient; dilatation of the sella tursica.

hypertrophy of the tissue of that gland. Is there degeneration of the cell elements of this organ or of its vessels, and what is its true nature?

Other pathological changes may also be regarded as fundamental, not only those which involve the skeleton, but some affecting other tissues and organs also. Among the latter may be noted the hypertrophy of the ganglions and nerves of the sympathetic (Henrot); persistence of the thymus; alterations in the thyroid body; hypertrophy of the heart and blood-vessels, which Klebs and Fritsche have drawn attention to; alteration in the organs of speech, and those also of generation, which have attracted the attention of Marie, Freund, and Verstraeten. Of those belonging to the skeleton the most curious is the great deformity and alteration of the pituitary fossa or sella tursica. On looking at the base of the skull it is noticed that the surfaces of the three main concavities are more or less irregular, on account of abnormal projection of the ridges of bone which separate them. That which strikes most attention, however, is the great size of the central depression at the base of the skull. The sella tursica is increased in all its measurements—in length, width, and depth. In this pathological process are involved the sphenoidal sinus, the groove for the optic com-

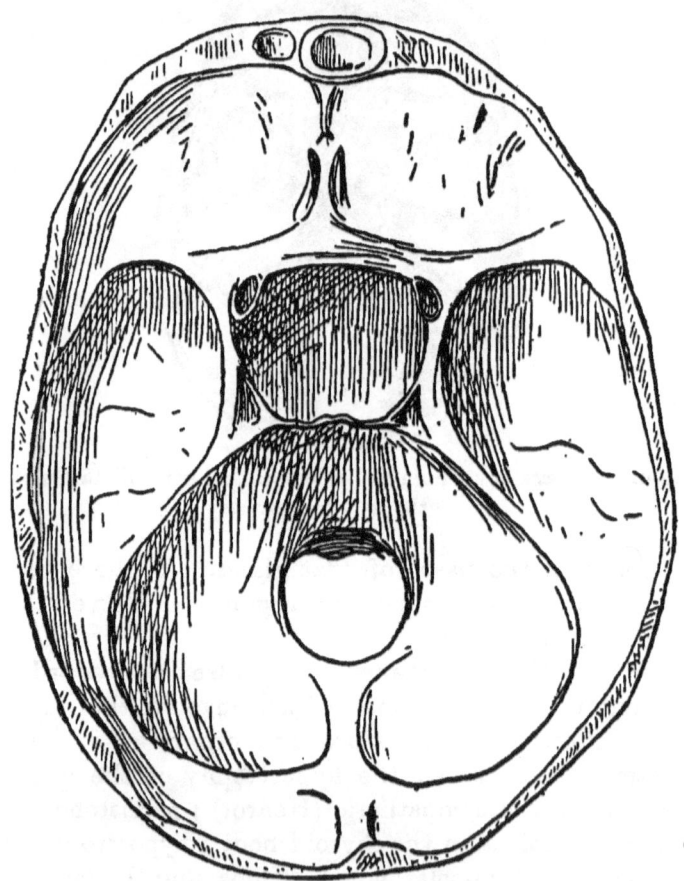

Fig. 17.—Base of cranium in its recent state of the patient who is the subject of Case VI.

missure, sometimes the cribriform plate of the ethmoid. The frontal, sphenoid, ethmoid, mastoid, and especially the maxillary sinuses (antrum of Highmore), are always more or less dilated, and their walls thinned. The vertebræ are hyper-trophied, and the depth of their bodies often reduced anteriorly, corresponding to the different curvatures. The clavicles and the iliac bones are hypertrophied, the ribs also

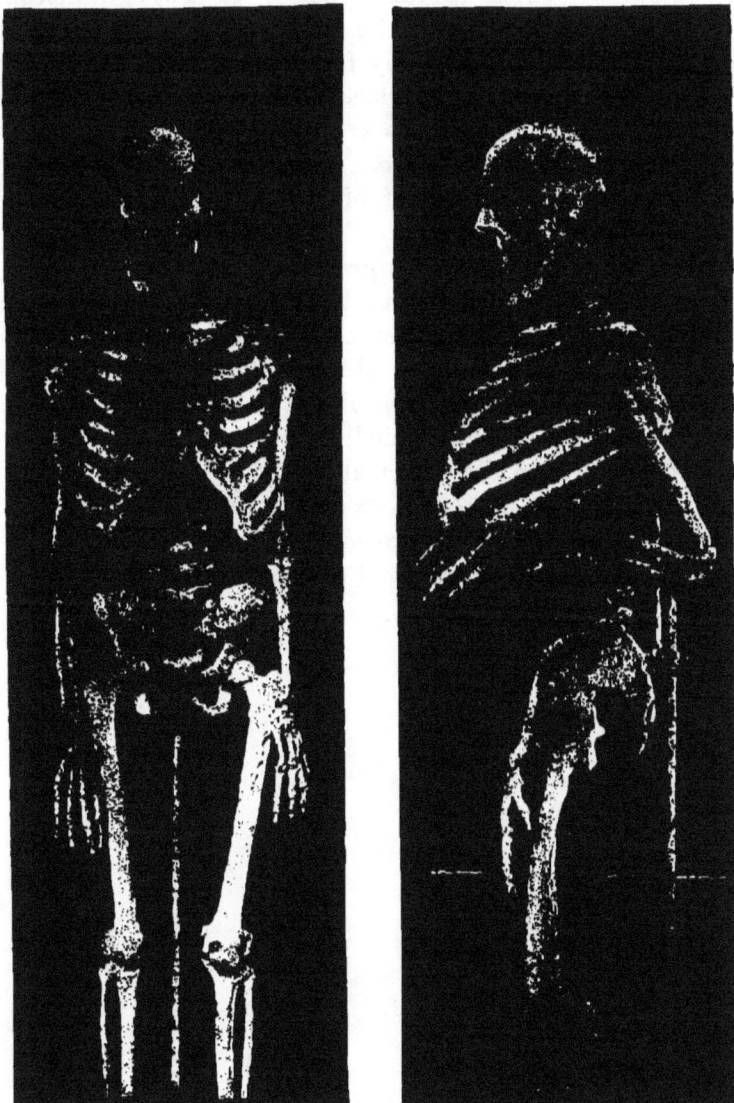

FIGS. 18 and 19.—Skeleton described by Dr. Taruffi. Very marked hypertrophy of the lower jaw and ribs.

are thickened and lengthened. As regards the long bones of the limbs, the most noticeable point is the increase in size of the ridges, lines, and projections for the insertion of muscles and ligaments. Their shafts, however, are neither lengthened nor enlarged, which is the opposite of Paget's disease. Finally, the short bones of the hand and foot are much increased in size. All the bony foramina for the entrance of vessels or nerves are more marked than usual; the spaces traversing the spongy tissues are specially increased in size. Broca has confirmed this latter point in an examination of the skeleton of a patient of Marie's, in which the spongy tissue in the diploë of most of the flat bones, and also in many of the long bones, was greatly developed.

We may note finally also that certain cranial nerves, particularly the oculo-motor nerve, may be increased in size; that the pineal gland also in Henrot's case was doubled in size. Certain of the viscera, either thoracic or abdominal, may be equally hypertrophied, and may have undergone marked changes in their normal relations.

Little is known as to the exact nature of acromegaly, or of the evolution of the process of which the deformities which attend it are the clinical outcome.

CHAPTER VI.

DIAGNOSIS.

[We have now enumerated and described the symptoms and characters which allow of Marie's malady being recognised easily and even at a distance. As, however, acromegaly has been confounded with other affections, and especially with certain possibly allied conditions, bearing features of resemblance to it, we will now review these analogies, and more especially the differences which serve practically to distinguish between them.]

BEFORE the appearance of the first memoir by Dr. P. Marie in 1886, a certain number of cases of acromegaly had been classed under such headings as Basedow's disease, cachexia pachydermia (Henrot, Wadsworth, &c.), osteitis deformans of Paget, leontiasis ossea of Virchow, gigantism. They had been confused also with certain other osseous lesions, such as more or less general hyperostosis of a rheumatic nature. Since the above-mentioned monograph was written the opposite mistake has been committed, and patients have been classed as the subjects of acromegaly who were suffering from some totally different malady. Among these cases have been certain forms of chronic rheumatism, morbid conditions dependent on rickets, and certain affections of the respiratory apparatus. This latter source of fallacy merits special attention, as will be shown later.

Myxœdema or Cachexia Pachydermia (Charcot).

In the subjects of myxœdema we observe, it is true, an increase in the size of the body; but this consists in a swelling limited to soft structures, which at once distinguishes it from acromegaly. In the former the skin is involved in a solid œdema, of a yellow or waxy colour; there is desquamation, and what is of still greater importance, the skin becomes adherent to the subcutaneous tissue. It is from this adhesion that the waxy and immoveable condition of skin results. In the latter the skin preserves its mobility hardly impaired, or acquires sometimes, as we have seen, a greater mobility than normal. The face in the former is puffed out and rounded in form, "in full moon" (Sir William Gull); the forehead is bossy; the œdematous lids can be opened only with difficulty, and almost cover the eyes; there is no prognathous condition of jaw; the hands and feet are swollen, and the limbs may be also swollen and rounded. In the second disease the face is not round; it is lengthened on account of the deformed hypertrophy of the lower jaw and chin, which produces a very marked prognathous condition. It presents an elongated oval from above downwards, with a low forehead. The hypertrophy of the extremities involves not only the soft structures, but the skeleton also is deformed. In myxœdema there are none of the curvatures, nor the double hump, found in acromegaly. Finally, the thyroid body is generally atrophied in the former; also the mental condition differs widely in the two diseases. Error in diagnosis is consequently, as a rule, easily avoided.

With regard to the above-mentioned disease, it may be remarked that recently Dr. Cheadle has endeavoured to class acromegaly as a variety of sporadic or endemic cretinism, strumous cachexia, and surgical myxœdema. Dr. Cheadle's paper, published in the 'British Medical Journal,' does not give the reasons for advancing this hypothesis, and we fail to see the links which connect such different maladies with varieties of cretinism. It is no doubt true that myxœdema and acromegaly are both diseases which affect most of the tissues of the body; but it is also not the less the fact that the principal anatomical alterations of acromegaly are very different from those of myxœdema. Among cases of the latter disease there is never present the remarkable swelling of the pituitary gland, with its accompanying lesions (of sight, &c.); nor the characteristic alteration in the skeleton, notably that of the hands and feet.[1]

---

[1] The diagnosis between osteitis deformans, leontiasis ossea, and acromegaly is here discussed, but it is mainly given also in Marie's original paper (q. v.).—Tr.

Chronic rheumatism, or more precisely certain changes produced by it in the extremities of the limbs, may resemble the features of acromegaly; particularly certain marked deformities of the fingers and joints. There are, however, these marked differences: pain is a much more marked symptom than in acromegaly, and predominates in the joints, which show from time to time characteristic signs of inflammation. There may also be present the distinct cracking sounds of chronic arthritis, with some functional disturbance. In short, the different progress of the two maladies serves to differentiate them definitely. In his memoir, published last year in the 'Progrès Médical,' Marie alludes to another possible error of diagnosis, somewhat more difficult to eradicate than the preceding. There exist certain individuals who combine the manifestations of rickets and lymphatic disease. The symptoms which confuse them with acromegaly are the following:—a more or less marked increase in the hands and feet, a thickening of the lower lip, causing it to be everted; sometimes swelling of the face, and a certain amount of palpebral œdema. A closer study of the hands and feet, however, cannot fail to establish the diagnosis, for, as Guinon has observed, the fingers are distinctly nodular, the hand is bony and deformed, and the soft parts do not present the enormous rounded prominences so characteristic of acromegaly. There is no cervico-dorsal curvature, no macroglossia nor prognathous jaw; and finally, the evolution of its symptoms differs from that of acromegaly.

Gigantism has been confounded with acromegaly by more than one author. At first sight without careful examination the mistake is easily understood. Comparison, however, between the state of the ends of the limbs of these individuals, especially the want of proportion between the size of their extremities and their length in acromegaly, quickly demonstrates an important difference. Two other groups of facts also prove that gigantism is not the same as acromegaly. These, as M. Guinon has stated, are first: "The subjects of acromegaly, as Dr. P. Marie has already demonstrated, are far from being giants. It is true that, allowing for the diminution of height due to the kyphosis, certain patients may attain 6 ft., yet the greater number are less than 5 ft. 9 in.;

and in women a height of 5 ft. 3 in. would include the majority of cases. So that though it must be allowed that occasionally the height in acromegaly is above the average, yet there is no question that gigantism is not one of the essential features of this malady. It may also be remarked that the heights of patients are greater in the recorded cases of former times than in the recent ones; the attention of observers having only been attracted to those persons whose increased stature made the colossal size of the limbs more noticeable." To this we may add a point of not less importance, that frequently in acromegaly the height, as well as the weight, after being increased in such a manner as to suggest gigantism, subsequently diminishes in a not less notable degree. "Secondly, in gigantism the appearance is never that which exists in acromegaly; the different proportions of the limbs and face preserve their relations to one another. Finally, the age at which these changes in the size of the limbs appear, their progressive increase, all show that they are not only absolutely different conditions, but that, contrary to gigantism, which is often only an exaggeration of a normal process, acromegaly is a true disease."

Another morbid condition, which has not received sufficient attention from authors, has often been confounded with acromegaly. It presents, however, considerable differences, as we shall presently show. The patients in whom it has been observed have been specially studied by M. P. Marie. The latter, commenting on this addition to pathology,[1] groups them "not as a separate morbid process, but as a secondary phenomenon; constituting, so to speak, an accidental complication in the course of an already existing disease."

Before giving the chief distinctive features of this affection we may briefly mention its history. Under the term "general hyperostosis of the skeleton," Friedreich published in 1868 an incomplete account of two patients, the subjects of a considerable hypertrophy of the bones; more especially of the hands and feet. These patients were the two brothers Hagner. Their disease was differently classified by authors, some regarding them as examples of Paget's disease; and others, among whom was Professor Erb, considering them

[1] 'Revue de Médecin,' Jan. 10th, 1890.

cases of acromegaly. Finally M. P. Marie showed, at a lecture given a month ago at the Salpêtrière,[1] the patient whom he had studied in the practice of M. X. Gouraud, and who had furnished him the opportunity of describing hypertrophic pulmonary osteo-arthropathy. Taking the characters of this group of symptoms, and beginning with the head, we find that the face does not present the long oval of acromegaly; that the nose may, it is true, be a little large and red, as if from excessive drinking (Gouraud-Marie case), but that it never has the dimensions of the nose of acromegaly. The lower jaw does not present any increase in size, nor the abnormal widening of its angle so constant in acromegaly. The chin is not thickened, and there is no prognathism—a difference of considerable importance from a diagnostic point of view. The upper jaw is deformed; "this deformity consisting in a marked thickening of its alveolar border and its posterior ends, giving a distinct slant to the palatine arch." The lips, tongue, neck, and larynx are not increased in size. The neck, contrary to what exists in Marie's disease, is never enlarged, and may be diminished.

The curvature of the spine is lumbar and lower dorsal, differing from that of acromegaly, which is cervical and upper dorsal. When two of these patients are compared together, the difference between the two curvatures is at once evident. This kyphosis, which is more than a coincidence, since Marie has observed it at least four times in eight cases, is not, however, constant in the disease in question; whereas it is so in acromegaly. "It should also be noted," Marie states (p. 30), "that it constitutes often a late phenomenon, not appearing till many years after the onset of the other bony deformities. The whole deformity of the thorax, resulting from the curvatures, differs from that of acromegaly. Occasionally it presents a triangular pyramidal form (Gouraud-Marie case), being unsymmetrical. The retro-sternal dulness may be more frequently present than in acromegaly. It was first observed by Erb in the brothers Hagner, and has since been noted by Ewald, and also, in a much greater

[1] "A Clinical Lecture on Acromegaly," in the 'Bulletin Médical,' December, 1889.

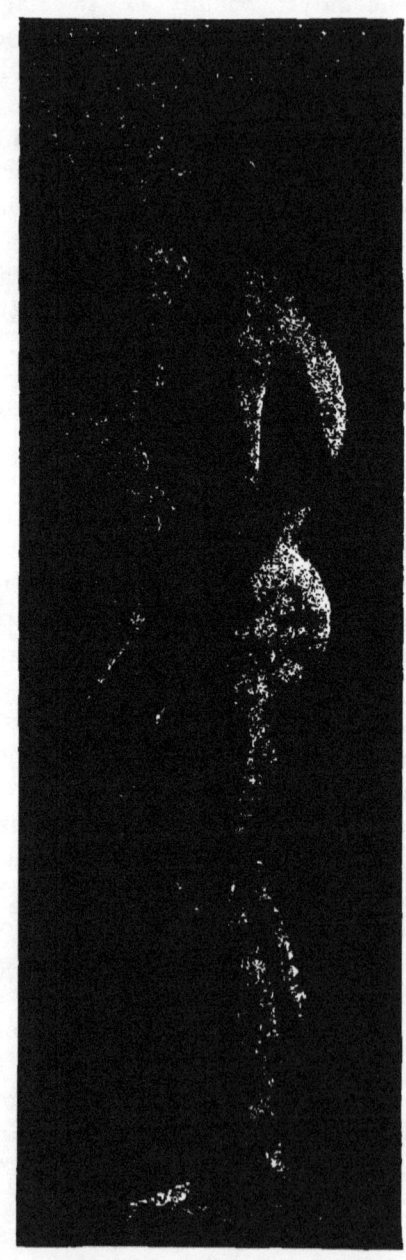

Fig. 20.—Patient the subject of hypertrophic pulmonary osteo-arthropathy. The kyphosis is distinctly lower dorsal.

DIAGNOSIS. 73

degree, by Gouraud and Marie. It has thus been recorded in three out of eight cases, and appears to be due to an increase in the bronchial glands. In the characters presented by the hands in the two maladies, the following alterations are found. "The hands," Marie states,[1] "are truly enormous; I may even state that they are more striking in their size than the hands of acromegaly. Whilst those of the latter are simply short, thick hands, those of hypertrophic pulmonary osteo-arthropathy attract attention not only from their size, but also from their deformity. The three divisions of

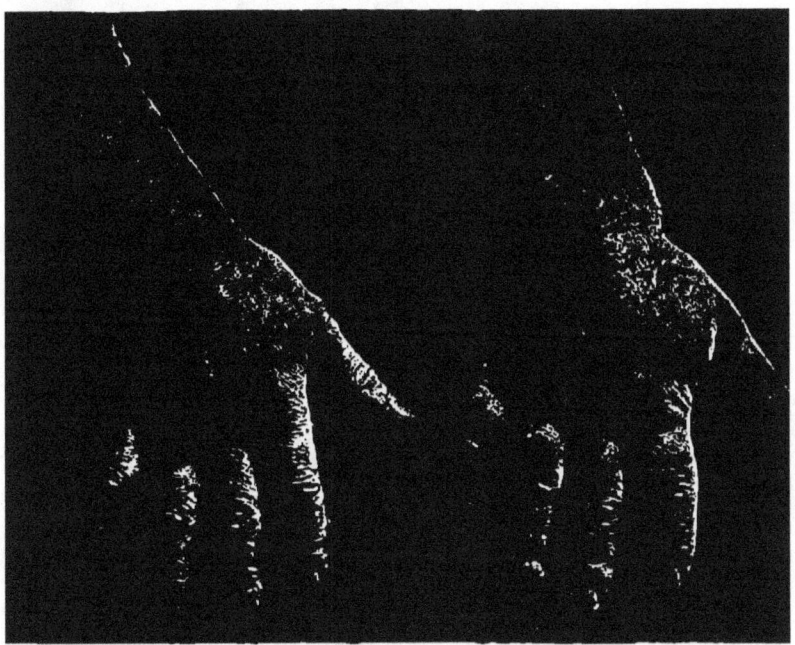

FIG. 21.—Hand of a patient the subject of pulmonary osteo-arthropathy compared with the normal hand of a person whose height is 1·75 m.

the hand should be separately studied—the fingers, the hand proper (the carpo-metacarpal region), and the wrist.

"Of these three segments the fingers present the greatest changes. They may be a little lengthened, but in all cases are considerably enlarged. These dimensions exceed those of the fingers of acromegaly. The size of the fingers is

[1] Loc. cit.

specially noticeable at the last phalanx. The latter is considerably swollen and bulbous, to such an extent that both relatively and actually it is the most hypertrophied of the three. The fingers also present the shape of the legs of a stool; which never occurs in acromegaly, in which the form of the different segments is well preserved. The nail at the end of the deformed finger also shares in this deformity, being considerably widened, lengthened, and more especially curved. This takes place to such an extent that the thumb of our patient, seen sideways, has a resemblance to the head of a parrot with its curved beak. Nothing at all similar to this is observed in acromegaly, in which the nails appear too small for the large phalanges which they cover.

"The very definite longitudinal striation of the nails must also be noted, and the marked tendency they show to become cracked and to split in their long diameter. Their peripheral extremity shows fairly often a very marked rosy colour. In three cases an abnormal tendency of the last phalanx to become hyper-extended has also been noticed.

"As regards the hand proper (the carpo-metacarpal region), it differs in a contrary manner; it does not depart perceptibly from the dimensions or form of the normal hand, beyond a little hypertrophy of the heads of the metacarpals. The contrast between the size of this part and the fingers is also striking. On the other hand, in acromegaly the metacarpo-phalangeal region is enormous; its increase in size may be said to be relatively greater than that of the fingers; it is this which gives the hands their battledore appearance.

"We now come to the third segment, the wrist. In all cases published at present, enlargement of this region has been noticed. In the various figures added to the cases this enlargement attracts attention. The lower ends of the two bones of the forearm are seen to be enlarged in an abrupt manner, and form an enormous projection above the hand. This thickening takes place as much in the antero-posterior diameter as in the lateral, and the lower portion of the forearm is seen to be as large as at the middle, or just below the elbow.

"It is also observed that the wrists are distinctly deformed. There is in acromegaly nothing resembling this. If, as happens in certain cases, the wrist is larger than normal, its size is still proportional to the rest of the upper limb; there is no projection, abrupt enlargement, nor deformity; the increase is, moreover, never to the same degree.

"Such is the appearance of the hand in hypertrophic pulmonary osteo-arthropathy, and that of the feet is analogous. If the latter are divided into three equal segments (the toes, the tarso-metatarsal region, and the ankle), it is found that the toes are, like the fingers, a little longer, but more especially enlarged. It is the last phalanx which is most in-

Fig. 22.—Hand of a patient of Dr. Saundby, the subject of pulmonary osteo-arthropathy; the wrist is deformed.

volved. It is the second segment also (the tarso-metatarsal region), as in the hands, which has shared least in the hypertrophy of the foot. It is specially opposite the heads of the metatarsals that the latter is most enlarged.

"As regards the third segment (the ankle), it is, like the wrist, enormously thickened in all its dimensions, and appears distinctly deformed.

"The lower part of the leg is found as large as the middle portion, bearing comparison with the foot of an elephant, recorded in several cases.

"In short, the foot presents absolutely the same characters as the hand; the only difference is that the increase in size is a little less marked in the former than in the latter.

"In this hypertrophy the principal share undoubtedly belongs to the bones. In acromegaly, on the contrary, the

Fig. 24.—Leg and foot of P. Marie's patient, the subject of pulmonary osteo-arthropathy; the malleolar region is much deformed.

soft parts are themselves hypertrophied in a corresponding degree to the osseous structures; it consists, in fact, of an hypertrophy *en masse* of all the tissues of the region.

"All the long bones of the limbs have their shafts, and

more particularly their articular ends, distinctly increased in size. This increase in their dimensions seems symmetrical in the two bones forming the same segment of a limb. The long bones of the lower segments (ulna, radius, tibia, fibula) are relatively more hypertrophied than in the upper portions of the limbs (humerus, femur). Besides the above-mentioned joints, which have been found enlarged, the same also occurs in the elbows and knees, but to a slighter extent. The patellæ are also equally enlarged. The above-mentioned kyphosis is probably also dependent in some degree on lesions of the intervertebral articulations.

"Besides the increase and deformity in the articular extremities, the joints present another important feature, *i. e.* limitation in their active and passive movements. These patients are, in fact, particularly clumsy with their hands, and their elbows present a more or less permanent degree of flexion. Their complete extension is impossible. The same also with the knees. It may possibly be said of the hip-joints that a somewhat analogous condition is present; moreover, in the shoulders of our patient (Gouraud-Marie) the movements were certainly a little impaired."

We have little to add to this description of Marie's, nor to his comparison of the important differences existing between acromegaly and hypertrophic pulmonary osteo-arthritis. The description above given explains at the same time the use of the terms in the name hypertrophic pulmonary osteo-arthropathy. As regards the prefix of the word pulmonary, it will be understood when we state that in eight of the cases contained in the above-mentioned work, two patients had empyema (Sollier and Lewis), two had neoplastic affections of the lungs (Ewald and Saundby), and three had different bronchitic lesions (Friedreich-Erb, Fraentzel, and Gouraud-Marie). Basing his conclusions on the knowledge of pseudo-rheumatisms, for the pathology of which we are indebted to Professor Bouchard, Marie includes the pathology of hypertrophic pulmonary osteo-arthropathy under the following headings :

1st. A lesion of the respiratory apparatus allowing, probably through the influence of micro-organisms, the introduction to these parts of septic matter (bronchitis, empyema, &c.).

2nd. Absorption and introduction into the general circulation of these products of the lungs.

3rd. Special morbid action of these substances on certain parts of the bones and joints, producing the lesions of hypertrophic osteo-arthritis. The latter morbid process presents no great improbability when it is remembered precisely how, by an almost analogous process, gout attacks with uric acid always or nearly always the same parts of the osseo-fibrous system.

There still remains one point as regards diagnosis which should be noted.

Certain cases occasionally occur presenting what some authors have incorrectly termed partial acromegaly. What is found in regard to these individuals? A considerable hypertrophy of the whole of one half of the body, which may be homonymous or crossed (a lower limb of one side and the upper limb and side of the face on the other). It may involve one hand or one foot, the fingers or toes, or sometimes one finger and one toe. These changes are accompanied by true deformities, and are always congenital; they remain stationary, and are most frequently unilateral. It is only necessary to recall some of the principal features of acromegaly to exclude these cases from the same category. "This term partial acromegaly," Guinon writes, "can only be a source of confusion, acromegaly being, on the contrary, as we have definitely shown above, an essentially general malady."

I will conclude the differential diagnosis of acromegaly by alluding to a disease which may, under certain clinical conditions, at first sight cause an error in diagnosis. This affection is vaso-motor paralysis of the extremities. This mistake was made in regard to a patient (Miss Erst, aged forty-four) who presented certain phenomena, particularly in the hands and feet, which led to the error.

What are the symptoms and modifications to which we allude? A marked increase in the size of the hands and fingers; a reddish tint with some livid spots on their dorsal surface; the occasional presence of pain in these parts; and a certain degree of duskiness extending up the arm, and even to the shoulder. Finally, the occasional presence of a slight degree of kyphosis in the upper part of the spine. These symptoms are, however, very different from those observed in the hands in acromegaly. In fact, the hands are not "camard," much less " battledore." The fingers are larger at their bases than at their extremities, which is not the case in acromegaly, in which they become almost as large at their extremities as at their base (" sausage fingers "). The nails have none of the characters of acromegaly, but are normal. The same distinctions apply also to the feet.

If we compare the cephalic extremity in these two maladies, we find, in the first, absence of the oval face and prognathism. The lips are not hypertrophied, but somewhat thin; the tongue is also not enlarged. The nose is aquiline, and presents none of the changes found in acromegaly. The cheek-bones, orbital ridges, and forehead are normal. Neither the speech nor voice presents the features of acromegaly. The spinal curvature is not cervical and upper dorsal, but involves the upper two thirds of the vertebral column, and is only slightly marked. The thorax is not deformed. The bones of the trunk are neither thickened nor enlarged; the same applies also to the long bones and joints. There is some migraine, but no constant headache. Vision and other special senses are normal. The neck is neither large nor short, as is found in acromegaly. This is sufficient, we consider, to differentiate between vaso-motor paralysis of the extremities and acromegaly. To enter further into the symptoms characterising the disease in question and its accompanying manifestations would be beyond the scope of the present work. We have limited ourselves to enumerating the three unusual phenomena which have been mentioned in classical works. These are hypertrophy of the hands and feet, sufficiently marked to suggest acromegaly; cutaneous insensibility, most noticeable in the end segments of the limbs; and a very distinctly diminished perception in the patient as to the exact position these parts occupy.

Treatment.

The treatment of acromegaly is but limited. It is confined to the relief of the worst symptoms, such as the headache, abdominal pains, insomnia, &c. Antipyrin very definitely relieves the headache. Great relief has also been experienced by the use of valerianate of caffeine. The glycosuria seems to be readily influenced by treatment (arsenic, alkalies, and dieting), at least in Marie's case. Knowing so little as to the etiology of the affection, it is impossible at present to formulate any very rational treatment.

CASES.

CASES.

CASE I (unedited, P. Marie and S. Leite).—M. X—, aged 36, single, a landlord, was sent to M. P. Marie by Dr. Hyberd (of Meung-on-Loire).

No hereditary antecedents of importance. His father suffered from "dead fingers." M. X— weighed 11 lbs. at birth. At the commencement of adolescence, while in Germany, he had acute rheumatism, which was worst on the right side of the body. Some time after he also suffered, like his father, from "dead fingers." Many years later, while in Denmark, he had another return of joint rheumatism, more severe than formerly. Never had syphilis. Between the ages of fifteen and twenty-three he went in for gymnastics, and became very strong. Between 1874 and 1875 (aged twenty-one) his parents and friends noticed an increase in the size of his hands, which took place little by little for some months. This hypertrophy, since his attention has been drawn to it, has been carefully watched by our patient, who states that its progress has been very slow but uninterrupted. The hypertrophy did not invade his feet for a long time. Three years after, in 1877-8 (aged twenty-four), all his features became more prominent; notably in the lower part of his face, which began to lengthen. At this time he had pains in his stomach, with disagreeable eructations, which have since persisted. Some of his teeth became carious and were extracted. A year later the patient began to stoop. At the beginning of 1881 one of his oldest friends failed to recognise him, on account of his "round back" and elongated face. He thinks that about this time his frontal region began to enlarge, particularly on the right side. At twenty-eight (1882), pains in the head appeared, which were almost constant, and were only relieved by taking food. During the following years all these changes which we have mentioned continued to progress, though in a slow manner; the cephalalgia deserves special mention in regard to this.

In October, 1889, Professor Bouchard made the diagnosis of acromegaly.

Present state (November, 1889).—Cephalic extremity: The cranium presents no deformity in its contour; it is small relatively to the face. The outer occipital protuberance and ridges are thickened. The mastoids are somewhat rounded. The frontal eminences are prominent. His hair is normally abundant. His face, on the other hand, is much changed. First, it is lengthened in a striking manner in the vertical diameter, being also a little widened. It is also slightly concave, which will be shortly alluded to. The forehead is a little retreating, on account probably of the projection of the upper part of the face. The patient insists that there is swelling of the right side of his forehead, but in reality the right frontal region seems only slightly thicker than the left. The orbital borders project, being thickened on inspection and to the touch. The lids are enlarged and very slightly pigmented, particularly so if comparison is made with that of the inner surface of the arm. Generally speaking, the colour of the skin is olive. His eyes project in a marked manner, but the lids can close completely over them. The cheek-bones and zygomas form distinct projections, noticeable at a distance.

The nose is thickened, especially in width; it is flat rather than prominent, and is particularly increased in size at its tip. It is a little deviated to the right side of his face, which gives it a slight asymmetry. Its alæ and septum are more thickened and resistant to the touch than normal. His lips are hypertrophied, principally the lower, which though not overhanging is beginning to be everted. His teeth are most of them bad; the lower molars also do not correspond exactly to the upper, the upper alveolar border coming inside the lower, on account of the thickening of the latter. The tongue is increased in size, especially in width; its papillæ, &c., are more marked. The palate is a little flat; its arch, &c., are certainly thickened. The tonsils are slightly enlarged, and the pharynx is manifestly thickened. The lower jaw is increased in size, more in thickness than length. The angle formed by its two alæ is wider than normal. This bone projects distinctly in front of the upper jaw, being inclined downwards and forwards. Resulting from this is the prognathism of the lower jaw, which would appear still more prominent were it not that the soft structures of the chin, together with the submaxillary glands, are also larger than normal. M. X— no longer resembles his photographs of nine years ago. He complains of headache, sometimes violent, during which he feels a desire to eat, and for which he is obliged to take meals more often than a person in good health. Moreover, if his hunger is relieved as soon as he experiences it, the headache disappears rapidly. "These pains increase," he states, "when the stomach feels empty." The headache, situated more frequently on the right side, is provoked if he leans his head against the back of a chair; it is increased by the warmth of bed, causing very distressing sleeplessness. He complains also of orbital and intra-ocular pains, coming on when he reads or fixes his attention on anything. When he goes to the theatre he is obliged to rest his eyes occasionally, or otherwise his sight troubles him, and causes involuntary shedding of tears, accompanied by pain. This weeping and difficulty in looking at things are also provoked by headache. Finally, he feels occasionally as if there were a weight of lead on the head. Examination of the sight by Dr. Parinaud showed the following: O.D. with $+0.75 = \frac{5}{10}$; O.J. Em. V. $= \frac{5}{20}$. Slight papillary congestion. Distinct grey discoloration and indistinctness of the disc, but without characteristic infiltration. The patient hears well on both sides. His smell is more acute than formerly; and he notices odours where other people do not perceive them. Taste and cutaneous sensibility normal.

His neck is large and short. The thyroid is examined with difficulty. Larynx hypertrophied. The pomum Adami is more prominent than usual, voice and speech harsh, and a little guttural. The voice has somewhat lost its tone. M. X— had a tenor's voice formerly; but now in 1881 it has become deeper, and he can easily produce that of a barytone or even a bass. The thorax is large; its anteroposterior diameter is longer than normal, on account of the lengthening and flattening to a certain extent of the ribs. The latter are thickened and widened, as may be ascertained by the touch; the interosseous spaces are much narrowed, especially the lower. The chondro-sternal joints are slightly thickened; also the angles of the ribs. The lines corresponding to the first two segments of the sternum show definite crests. The latter slants downwards and outwards, and is enlarged. The clavicles, the spine of the scapula, and the acromions are examined more easily than normal. The vertebral column is deviated. There is well-marked cervical and upper dorsal kyphosis; also a slight degree of scoliosis, but no lordosis. A marked double hump results from this, like that of "Punch." The abdomen is large, markedly pendulous, presenting large folds corresponding to the ridges. The thoracic viscera are normal. The area of dulness, which Professor Erb has described (behind the upper part of the manubrium sterni, at the commencement of the first intercostal space), is present, but only to a slight extent. The liver is normal to palpation and percussion; that is to say, it is neither raised, painful, nor bossy on the surface. The stomach seems a little dilated. The patient has had cystitis and prostatitis accompanied by their ordinary symptoms. The scrotum and testicles are larger than normal. The penis is not of excessive dimensions.

The hands of M. X—, which take No. 9 gloves, are truly enormous, both in width and thickness. Their length one may state is normal, and the disproportion between their dimensions is very striking. All the soft structures are thickened and enlarged to a very considerable degree. The folds and depressions opposite the joints are in consequence much increased. The fingers are very large, and as much so at their extremities as at their base. The hands present the "battledore" appearance. The patient experiences the sensation of dead fingers when the atmosphere is damp. The nails, which are small and flattened, are out of proportion to the fingers; they have preserved their normal structure, and are not cracked, but show a longitudinal striation.

Though his wrists are a little increased in size, they bear no proportion to the hands. His forearms and arms may possibly be a little enlarged, but it must be remembered that M. X— has always been muscular. The joints are not enlarged to an appreciable extent. The lower limbs are not the site of any appreciable deformity; they are large, but not excessively so. Some of the veins are a little dilated on the limbs, on the skin of which there are some small patches of eczema. The joints are nearly normal, except for some gratings in the knees, as also in the joint of one shoulder. The patella reflex is diminished. The ankle and malleolar regions are enlarged, but to a slight extent. The feet are thickened, widened, and almost flat-footed. The changes are identical to those of the hands, only to a somewhat less marked degree. Among the toes, the first three are the largest; their nails are flattened and longitudinally striated.

It is difficult to say exactly whether the patient has an excessive appetite, for though he eats several meals a day, these are not excessive; and he is obliged, moreover, to take nourishment in order to relieve the headache and ocular pains. He takes fluids well, but it is necessary to drink by small gulps, as large quantities at a time cause vomiting. There is tendency to slight hæmorrhoids and constipation.

The movements of all the joints are executed normally, except in the metacarpo-phalangeal and phalangeal joints, which are incapacitated by the enormous folds described above. Mental faculties normal. Some vertigo, without direct relation to the headache. The sexual function is a little diminished.

The skin has an olive tint, more marked on the face and lids; it is thickened, frequently itches, and perspires abundantly. There are six or seven molluscum fibrosum tubercles on the right side of the neck, and one large one on the inner side of the right armpit.

His urine was examined two years ago by his doctor, and gave a negative result as regards sugar and albumen.

The following is the note of Professor Bouchard, who first discovered peptones in the urine of this patient:—" Peptonuria, a large precipitate with Tanret's reaction, soluble in heat. No deposit with picric acid. Abundant precipitate with absolute alcohol. Intense violet with Fehling's solution, used cold. No albumen nor sugar. A mahogany tint with $FeCl_3$" (October 23rd, 1889).[1]

[1] The detailed measurements of this patient are here given.—TR.

CASE II (P. Marie and S. Leite). (Unpublished.)—M. Z—, aged 39, a Doctor of Medicine, of Jewish birth, was recognised by Marie, while in a tramway, as the subject of acromegaly. His mother died of interstitial nephritis; her polyuria was specially marked at night. M. Z—, who gives no other hereditary antecedents, has always been sober, and has not had syphilis. M. Z— asserts that while he was studying medicine in Paris his comrades used to notice that he was big and strong for his age, which makes him think that possibly his malady had already commenced at that time, namely, between twenty-two and twenty-six. It was, however, later than that, between 1879 and 1880, at about the age of twenty-eight, that he noticed that the fingers of both his hands had begun to increase in size; and that the size of his gloves, which was $7\frac{1}{2}$, had become too small. At this time M. Z— had already suffered from more or less severe attacks of bronchitis, and from erratic pains in the limbs.

In 1881, after having had recourse to various treatment without distinct benefit, he resolved to go through a course of hydropathy, and specially of hot baths. Instead of amelioration, this resulted in more severe attacks of bronchitis, attacks of coughing of long duration, and pain in the upper part of the chest, notably in the precordial region. In the latter they were sufficiently severe to suggest angina pectoris. It may be stated that M. Z— was rather a hypochondriac. At this time (1882-3) he had intense thirst and a great appetite. The increase in the size of the hands and fat progressed very slowly, but in a definite manner; his face also began to alter, principally the chin, lips, tongue, and nose. As regards the tongue, M. Z— stated that it caused him inconvenience for talking, mastication, and swallowing, and that he had consulted several doctors for its enlargement.

Present state (commencement of November, 1889).—On the day when M. P. Marie encountered M. Z— on the tramway, the appearances which struck him most were the oval lengthening of the face, the prognathism at the lower part of it, and the marked spinal deformity.

Cephalic extremity.—His hair is abundant, though there is a slight extent of baldness. The hairs are of normal size, and are beginning to turn grey. No thickening along the cranial sutures. Mastoid processes normal. External occipital protuberance much developed, as well as the curved lines radiating from it. The cranium taken as a whole is not deformed. His hats had been enlarged during his malady. His face has a distinctly oval form from above downwards, but is not asymmetrical. The forehead is low. The orbital ridges are not more thickened than normal. The eyebrows are well developed. The lids are not thickened, but the skin covering them has a browner pigment than over the forehead. The eyes are small relatively to the size of the face; no exophthalmos. Movements of the globes normal. No intra-ocular pains. The usual projection of the cheeks is exaggerated. The nose is large and a little flattened; the septum and alæ of the nose are thickened and rigid. The skin of the face is a little pale, and of a dull tint.

The dimensions of the mouth are a little increased. The lower lip is distinctly thicker than the upper, but it is not overhanging. The teeth are more or less carious, and those of the lower jaw are separated from one another. The tongue is increased in size, especially in thickness, and its normal fissures are more marked. The roof of the mouth is longer than normal. The uvula is hypertrophied; also the pillars of the fauces and tonsils, but to a less extent.

The chin is prominent, and the depression between the lower lip and the symphysis of the jaw is marked. The lower jaw is wider and longer than that of a man in good health; it is increased in all its dimensions: from this result the prognathism of the lower part of the face and the projection of the lower dental arch. The upper jaw is perhaps a little thicker than normal. The ears are slightly increased in size, being possibly thicker.

The neck is large, appears short, especially in front, on account of the curvature of the patient's spine. The larynx is certainly very thick and hypertrophied. At the time of undergoing the hydropathic treatment M. Z— had a severe laryngitis, which left a harsh deep voice. The hyoid bone is a little hypertrophied. The hands, which present their normal shape, are widened, thickened, but not longer than before the present malady. They are somewhat flat. The large furrows in the palms of the hands and the folds of the flexures corresponding to the metacarpo-phalangeal joints, and those of the phalanges are much more accentuated than those of a normal hand. The soft structures which separate the latter are correspondingly much more hypertrophied than in a normal hand, in such a manner that they form large pads, such as exist in all these patients, which can be seized between the fingers. The skin of the hand has a normal appearance, and presents none of the discoloration characteristic of myxœdema.

The fingers are very large, as much so at their tips as at their bases; some of them are curved, notably the middle finger. This curvature involves the last two phalanges. There are some nodosities on the last phalangeal joint. There is a synovial cyst under the tendons of the long abductor and extensor of the thumb, opposite the thumb-joint. The nails are of normal length, and are flattened. Their borders are curved; they are not increased in width, causing them to appear small in proportion to the rest of the thickened soft structures. There is slight longitudinal striation. No atrophy of the dorsal interossei. Four years ago M. Z— took $8\frac{1}{4}$ gloves, now $8\frac{3}{4}$.

His wrists are a little increased in size, but are in marked disproportion to the hands. The muscles of the forearm and arm are well developed. The movements of the fingers and hands are somewhat incommoded by the size of the soft structures of these parts, but their strength is practically normal. The joints of the upper limbs are normal, and there are no gratings in them. The extremities of their bones are not increased in any appreciable manner.

The trunk is the site of important changes. The right shoulder is more elevated than the left. When he was eight years old his tailor told him that one shoulder was a little higher than the other. On examining the spine one is struck with the marked hump extending from the last cervical vertebræ to the ninth or tenth dorsal. It presents also a less marked dorsal scoliosis, with convexity to the right. The lumbar curve is more accentuated than normal, possibly to compensate for the hump. The thorax presents slight lateral flattening, but its anterior portion is not notably projected forwards. The clavicles are a little large.

The ribs are thickened, but not excessively so. The angles of the ribs are more hypertrophied on the right than on the opposite side. The chondro-sternal joints form a marked projection. The junction of the first and second portions of the sternum forms a marked crest. The sternum is slightly oblique downwards and forwards. The side view of the abdomen presents nothing abnormal; it is a little pendulous. The hairs on the trunk are numerous, and specially developed. The area of dulness described by Erb is not present. Examination of the heart shows it to be normal. The whole chest is resonant, with signs of pulmonary emphysema. Percussion over the liver gives dulness to the level of the breast. Hairs of pubes normal. Testicles normal. No glandular enlargements. Slight hæmorrhoids. The pelvic limbs are practically normal in their shape and size. Their muscles are well developed, and all movements are well executed. The knees are only slightly thickened. There are slight varicose dilatations on the limbs. The articular ends of the tibia and fibula, especially their malleoli, are large, but immediately above them the limbs are normal. The feet are larger and thicker than normal, notably their soft structures; the same also with the toes. The hairs on the lower limbs, like those on the rest of the body, are much developed. Patella reflex normal. Tendo Achillis normal. Sight feeble. Hearing and taste normal. Cutaneous sensibility not impaired; the same with the intellectual faculties. The patient is easily fatigued with the least exercise. M. Z— has no increased appetite, but an excessive thirst. He has perspired much for the last five years; this has disappeared spontaneously. Polyuria at night; oxaluria (?). M. Z— states that he has never found albumen nor sugar, which corresponds to the analysis of Dr. Beaumetz. The skin has a slightly dark colour, and presents some tubercles of molluscum fibrosum opposite the nape of the neck.[1]

CASE III (Marie).—B—, æt. 49, a type-writer.[2] Nobody in his family has shown any increase in the size of his limbs.

[1] Detailed measurements given.
[2] This case was published by Marie in the July number of 'Brain,' 1889. Some additional notes are here given.

His father, aged seventy-three, is still living, and enjoys good health. The patient was puny at birth. At fifteen he had typhoid fever. He went through military service. He denies any venereal complaint. In 1881 he had an eruption of painful confluent pustules on the hairy part of the neck. They presented acute inflammation for four or five days, and prevented sleep. They dried up after a few days.

The patient could not say at what precise period his limbs had commenced to increase in size. He stated that he had not attached much importance to the changes. As far as he recollects, during his service in the army, which continued to the age of twenty-seven years (1867), he did not wear large boots. "The shoes which I wore were rather small than large for my height." In 1878 he noticed an increase in the size of his hands. It was at this time (aged thirty-eight) that the present changes began. It appeared certain then that the first symptoms of acromegaly made their appearance between 1867 (aged twenty-seven) and 1878 (aged thirty-eight). At thirty the weight of the patient had already increased a little, from 64 to 80 kilogrammes; but it was especially at the end of 1878 that his weight and proportions had manifestly increased. In December, 1887, he weighed 113 kilogrammes; a year later he went back to 106. His height (taken with bare feet) at his regiment was 1 metre 67.

Present condition (January, 1889).—The appearance of the patient is so characteristic that the diagnosis was made in the street. His hands and his feet are enormous and thick; also his fingers. The hypertrophy of the soft parts is so extensive that it is impossible for the patient to completely flex his fingers to his hand. The left forefinger shows a scar left by a whitlow which developed in February, 1888, in consequence of a splinter; since then the patient has experienced at this point a sensation of numbness.

The wrist is thickened—less so, however, than the hand; the forearms and arms are also large, but their dimensions do not, like the hands, give the idea of monstrosity.

The two segments of the thoracic limbs are proportionate in size; their muscles are well developed, without true hypertrophy. The patient used to be very robust, and ranked as very strong among the members of his corps; since then he has been weak for his age. The face is also characteristic; the nose is large and a little flattened. The upper lip is thick, but much less so than the lower lip. The latter is enormous, giving the appearance of an overhanging pad. The lower jaw is much hypertrophied. The chin projects strongly forwards, and the prognathism is such that the lower incisors project beyond the upper about 8 millimetres; the teeth are not enlarged or thickened. The tongue is very wide, long, and thick. Speech has become not exactly difficult, but a little thick; probably on account of the hypertrophy of the tongue and soft structures of the buccal cavity.

The eyelids and ears are normal. The hair is well developed; a little coarse, but this B— states has always been so. The beard is abundant and somewhat frizzly. The neck is thick and short. There is marked cervico-dorsal kyphosis, which produces an inclination of the head forwards. The larynx is large, but cannot be said to be hypertrophied. The thyroid body cannot be felt distinctly enough to say if it is normal. The voice is strong and a little deep in tone. The patient cannot sound musical notes.

The thorax is large and the sternum very oblique. The xiphoid appendix is very prominent and hypertrophied, forming a distinct projection under the skin. The sides of the thorax are a little flattened. The retro-sternal dulness of Erb is not present.

The heart is a little hypertrophied, its sounds are normal; the pulse is small and compressible.

The genital organs present nothing abnormal. The sexual function, which was never excessive, has diminished the last few years (it should not be forgotten that the patient is diabetic). The skin has the olive colour which has been found in some other patients; it is not, however, a clear tint. There is a tubercle of molluscum fibrosum on the right shoulder. The cutaneous sensibility and special senses are not altered. As regards the sight, however, we must make the exception that its acuteness is limited for small objects. Ophthalmoscopic examination was negative. The patellar reflexes are much diminished. There is slight but distinct dilatation of the veins of the legs, particularly over the internal malleoli.

The mental faculties are good, and he is more intelligent than the majority of individuals of his class, being a self-taught man. No headache.

The patient's appetite has been excessive, notably during the last three years; it is the same with the thirst—he drinks on an average five litres of wine. To quench his thirst he drinks much at a time.

Examination of the urine reveals the presence of a large quantity of sugar. We have thus a case of diabetes, and it is impossible to say whether the polyphagia, the polydipsia, and the polyuria have any relation to the acromegaly, or are, on the contrary, the consequence of diabetes.

May 15th, 1889.—In order to relieve the diabetes, he was prescribed in January alkalies, arsenic, and dieting; the sugar diminished very rapidly. No trace could now be discovered by the aid of the copper potassium solution. At the same time the thirst diminished. The patient is much more active, and feels much less tendency to become fatigued and sleepy than he did before the treatment.[1]

Additional note.—Sensibility and cutaneous reflexes normal. " No distinct lesions of the fundus of the eye, unless a slight degree of vascular congestion. No diminution of the visual field. Vision acute." (Dr. Parinaud.) Bronchial secretion normal and abundant. The tonsils are enlarged, also the uvula, which contribute between them to render pronunciation and deglutition difficult. He has lost a certain number of teeth. The ribs are lengthened and widened. The sternum is thick and enlarged, and slants obliquely downwards and forwards. The bones of the pelvis are thickened. The orbital borders are thickened. The larger joints and also the smaller ones may be stated to be normal. There is slight longitudinal striation of the nails. The temporal and radial arteries are somewhat tortuous. There are some bleeding piles.[2]

Analysis of urine, November 22nd, 1889:—" Reaction acid, sp. gr. 1014, no sugar, albumen 0·08, earthy phosphates

---

[1] Detailed measurements of the patient given.
[2] Measurements of head given.

1·0091, alkalies 1·15, total 2·1391, urea 0·8855, uric acid 0·205, no residue, urine limpid, colour pale yellow. Peptones to the reaction of Tanret and Millon (rose colour)."

Analysis of November 23rd : "Acid reaction, sp. gr. 1013, no sugar, albumen 0·153, earthy salts 1·02, urea 9·067, uric acid 0·217, no fixed residue, total 2·170; alkalies 1·15 ; colour, &c., as before. No deposit after standing twenty-four hours. Microscopic examination after thirty hours' standing : Pus globules, crystals of ammonio-magnesian phosphate and oxalate of lime. Diminution in the deposit of urate of soda. Crystals of cystine in regular six-sided prisms. Peptones to the same reactions."[1]

The results, however, of the analysis by Dr. Berlioz on the same patient were the following : " General characters normal. Excess of uric acid in proportion to urea, of which the quantity is slight ; trace of albumen ; some leucocytes and epithelial cells from the bladder ; no sugar nor biliary pigments ; no peptones. With Fehling's solution a simple yellow discoloration is obtained, without formation of precipitate. Negative results from polar examination. Tanret's reaction doubtful, but this test is insufficient to establish the presence of peptones. In fact, to find this substance the process of Hofmeister must be adopted, the reaction of biuret does not take place, and, therefore, no peptones are present."—February 3rd, 1890.

CASE IV (P. Marie) ('Nouv. Icogr. Photograph.,' 1888).— M. C—, æt. 45. His mother is eighty-two years of age, in excellent health, and has not lost a tooth. His father died at fifty-eight (cerebral softening ?). All his uncles and aunts were healthy, and no one has presented anything similar to the affection from which he suffers. The members of his family are in general a little above medium height, but not of any excessive size. He has had only one brother, who died young. He himself always enjoyed good health during infancy, but never had a good complexion. At twenty-three he had undoubted syphilis, with slight subsequent manifestations, except in the throat. Up to the age of sixteen the patient was of ordinary height, inclined to be short ; when

[1] Further details of Dr. Grenouillet's process of analysis given.

about this time he all at once began to grow. From seventeen to eighteen his height increased to a considerable extent, and since then, up to the age of forty, he is convinced that he has not ceased growing, though very little, it is true, yet definitely. At the age of nineteen or nineteen and a half, when leaving college, he did not measure less than 5 feet 8 inches; from seventeen to eighteen he had increased 11 inches, from eighteen to nineteen 4 or 5 inches; from the age of nineteen and a half to the age of forty years he had gained 3 inches; since then he measures 5 feet 11 inches.

The patient used to have good health and great muscular power; he was a soldier, and could stand fatigue well.

About the age of thirty-two or thirty-three he noticed (without being able to assign any reason) that his hands, which had, it is true, been always strong and muscular, had increased distinctly in size and were progressively enlarging. It seems as if this increase has not yet completely stopped, for lately the patient believes he has had more difficulty in putting on his gloves. The hypertrophy of the feet showed itself about the same time as that of the hands.[1]

The shape of the hands is in the main regular, but enlarged. The soft parts have undergone a considerable hypertrophy, which involves specially the palmar surfaces of the fingers, notably opposite the last phalanx and hypothenar eminence.

On the back of hands and posterior surface of the forearm there are marked venous plexuses. The size of the veins at these parts seems certainly larger than those of a healthy hand (it is possible that the difference in the thickness of the skin may account for part of the difference).

The nails are not very large in comparison; in fact, it is doubtful if they are at all hypertrophied; they probably appear large compared to a normal hand on account of being to a certain extent flattened and curved. They present a very marked longitudinal striation. Their length is not increased, but appears from the hypertrophy of the soft parts to be a little short.

On the palm of the hand the skin over the heads of the metacarpals is considerably increased in thickness.

[1] Detailed measurements of the hands are given.—Tr.

With regard to the rest of both the upper and lower limbs, there is no muscular atrophy, but the soft structures are in general somewhat flabby; the muscular power is of medium strength in proportion to the patient's stature. As regards the trunk, what immediately attracts attention is the curious forward projection of the thorax. It presents, in fact, the front hump of "Punch." This analogy is the more similar, as there exists also at the same time a certain degree of projection at the back.

The anterior part of the thorax forms an obliquely inclined plane from above downwards and from behind forwards, making an angle with the spine of about 45°. A considerable projection forwards of the false ribs results from this; at the upper part below the clavicles there is a distinct flattening.

From the lower cervical vertebræ to the level of the inferior border of the scapula there is a very distinct hump; below this is a depression partially corresponding, though not the same in degree, to the forward projection of the false ribs described above.

There is possibly a slight lateral curvature of the spine towards the right, but it is so slightly marked as to be hardly definite. The left shoulder is a little higher than the right.

On palpation the clavicles do not seem to present the increased dimensions which should belong to a man of his stature; they are possibly a little more oblique than in the majority of persons.

The form and dimensions of the scapulas cannot be determined with sufficient accuracy to say anything definite with regard to them. The ribs are considerably increased in size. The xiphoid appendix is very large. Respiration is diaphragmatic and lower costal in character. The patient walks well and can travel without difficulty, but becomes easily fatigued on mounting stairs. The pelvis itself does not seem to be increased in size. The thighs and buttocks present nothing particular; their muscles are a little flabby, and the skin is dry and rough. The testicles are large; the penis is distinctly larger and longer than normal. The hairs of the pubes are abundant. The patient states that there is diminished sexual desire.[1]

For four or five years he has gone to the same bootmaker, and during this time the latter has made his shoes the same measure and shape, so that it seems as if his feet had remained stationary during this time.

Over the whole surface of the body the skin has a mixed brownish-yellow colour, of a slightly olive tint; this is most marked on the face. The skin is flabby, and as if too large for the body; it is also a little thickened and dry. On the neck, shoulders, and back are several small tumours of molluscum pendulum.

On the legs some varicose veins are seen, but they are not much developed. There are also hæmorrhoids, which appeared about the age of twenty-five or twenty-six; these have often given rise to some bleeding. The patient has always had a good appetite, and eats perhaps rather more than the average, though not excessively in proportion to his size. His digestion is always good. He drinks a fair quantity of water at his meals, but never between them. He passes a fair quantity of urine; it is of a clear ordinary colour, and contains neither sugar nor albumen.

Face.—The patient had during infancy up till the age of twenty-two or twenty-four a round face and small nose; at the latter age it commenced to enlarge. It is, however, only since the age of thirty that it has been really deformed. The general shape of the face is much lengthened.[2]

The forehead is oblique; the superciliary ridges project considerably forwards; while behind the latter are depressions, which resemble this region in the cow. The eyebrows are well developed. The lids show a browner pigmentation than the rest of the body, but are not thickened. Their width measured across the lower part of the superciliary ridge is 25 millimetres.

The eyes project slightly, and are of normal size, with a somewhat vascular conjunctiva. The pupil is small; there is a slight degree of myopia. The eyelashes are normal. The nose is very large; its greatest width is 4 centimetres

[1] Detailed measurements of lower limbs given.—Tr.
[2] Measurements of face given.—Tr.

(distance between the outer surface of the two alæ of the nose).

The septum is a little deviated to the right. The shape of the nose is regular; its tip is more pointed than flat. The cheeks form a very marked projection. The mouth is not increased in its dimensions; but the lower lip is very thick, and forms a large pad.

The teeth are good; their size is normal, rather small than large. The teeth of the lower jaw, which formerly were behind those of the upper, are now in advance. He used formerly to be able to cut through horsehair for fishing with them; but for twelve or fourteen years he has been unable to do so.

His tongue is much increased in width; its length is also distinctly increased. Its upper surface presents a great number of ridges, but these are all rather exaggeration of the normal conditions.

The upper jaw presents nothing very marked; it is probably a little more developed than normal, but this is not very definite. The lower jaw is thicker, longer, and wider than normal; its angle is certainly widened.[1]

The ears are rather large, somewhat thickened, but do not present anything definitely abnormal.

The cranium (over the frontal region) does not show any marked deformity. The sutures are, however, more easily followed with the finger. The outer occipital protuberance, and the lines diverging from it, project considerably. The hair is abundant, a little grey; its size seems normal.

In the neck, the larynx forms a very marked projection, but not excessive compared to the stature of the patient; its antero-posterior diameters seem, however, increased. The hyoid bone appears also to be an abnormal size.

The thyroid body, which is felt only with difficulty, seems diminished in size; it may certainly be said that it is not increased.

The patient's voice, which in infancy was fairly high and could reach the highest notes, became deeper at puberty; and is now powerful, very deep, and sometimes rather muffled. He cannot now reach high notes.

[1] Measurements of jaw given.—TR.

As regards the organs of special sense we may note:

The sight is, the patient states, almost perfect; and notably when hunting he has nothing to complain about it. In fact, examination by M. Parinaud showed the vision to be but very slightly defective: $V. = \frac{5}{10}$, for both eyes. No muscular defect with either eye. With the ophthalmoscope, however, a double neuritis is detected; characterised by red coloration of the papilla, with infiltration of its edges and adjoining retina. The veins are dilated and tortuous.

Hearing, according to the patient, is somewhat deficient; he hears well, however, when spoken to even with a low voice.

Smell is fairly good. Nasal respiration somewhat obstructed.

Taste good.

CASE V[1] (P. Marie; 'Revue de Médecine,' 1886). Additional note (Souza-Leite, January 23rd, 1890).—The dimensions of the hands and feet are the same as five years ago. At the present time, Mdlle. Fuschs perspires profusely on the least exertion. There are constriction pains in the temples, and in the bones of the cranium, "as if the bones were being scraped." The cerebral pains are almost continuous, with exacerbations at times. "I feel," he states in describing them, "as if the head was being compressed from before backwards." The skin of the neck forms a large pad opposite the occipital protuberance. There is possibly some dulness to the left of the manubrium sterni. With regard to the digestive functions and appetite, meat disagrees, and she eats vegetables exclusively.[2] The patient's strength is much diminished.

The urine of Mdlle. F. was examined January 24th, 1890, by Dr. Berlioz, who has communicated the following analysis: "Colour deep yellow, transparent. Abundant flocculent deposit. Odour normal, no reaction, sp. gr. 1023, urea 14·50, uric acid 0·386, phosphoric acid 0·273, chloride of sodium 16·20, no albumen, no peptones, sugar 0·55. Occasional leucocytes and epithelial cells from the bladder seen

---

[1] This case is translated in Marie's original paper, for the first part of which the reader is referred to it.—TR.

[2] Detailed measurements of different parts here given.—TR.

under the microscope. There is thus a very small quantity of sugar, no serum, no globulin, nor peptones. The proportions of urea, uric acid, and phosphoric acid are less than normal.

"I have tested for peptone according to Hofmeister's process—that is to say, with phospho-tungstate of soda,—and have not been able to detect it. Other reactions, comprising that of Tanret, also do not give any traces of peptone. When these reactions seem to show the existence of this element in the urine it is necessary to confirm it by Hofmeister's process, which is the most certain of all."[1]

CASE VI (P. Marie; published in the 'Revue de Médecine,' 1886).[2]

CASES VII and VIII (P. Marie).—These occurred in an adult man and woman, who had been seen by Marie and two of his colleagues at the Paris Hospital. In these patients the skin of the face is more pigmented than normal, and is of a dull tint. The face is lengthened from above downwards; the orbital arches and cheeks are prominent; the nose is large and flat. The lips are thickened, especially the lower. The chin is prominent and thick, especially when singing. These patients are bent forwards, and have thick necks. Their hands are enormous, wide, and thickened. The fingers are similar to those of all four cases of acromegaly. The voice on singing is deep and metallic. Such are the symptoms sufficient to form a diagnosis.

CASE IX (Péchadre, 1, *résumé*).—This case occurred in a woman of 42, a doorkeeper, in the practice of Dr. Bouveret, in May, 1889.

Hereditary antecedents.—Father died of heart disease at sixty. Mother died from cerebral apoplexy at eighty-two. No nervous diseases in her uncles or aunts.

[1] Other analyses given, made by Dr. Berlioz, Marie, and Souza-Leite, showing the presence of an "alimentary glycosuria." Detailed measurements of the head by Dr. Manouvrier also recorded, demonstrating that the head was abnormally large, and of a somewhat masculine type.—TR.

[2] For the details of this case the reader is referred to the translation of Marie's original essay.—TR.

She has had eight brothers and sisters, of whom four are dead. The first, at twenty-seven, of pneumonia ; the second, at thirteen, of an unknown acute malady ; the last two when very young. The four others are healthy.

Personal antecedents.—Variola at four. Good health during infancy. Menstruation began at seventeen, which was regular, but somewhat painful. The patient was obliged each time to take to bed for a day or two. At this time she was tall and slender. Her face was rather thin, and her hands and feet of medium size. She married at thirty-four years of age. From about that time she grew somewhat stouter ; never pregnant; no alcoholism; no syphilis; no ague. Eight years ago she had rheumatism of the right shoulder, and probably a phlebitis of the left leg, of longer or shorter duration. She has always had a small goitre, which has never caused inconvenience ; one of her sisters also presents a small goitre.

History of present malady.—Seven years ago, at the end of 1882, without definite cause (beyond mental emotion and fatigue), "menstruation entirely and finally ceased." The abdomen became enlarged, and attained a very considerable size. Violent abdominal and lumbar pains then compelled the patient to cease work, and to take to her bed at intervals. She consulted a doctor, who thought it was pregnancy and advised waiting. The pains disappeared of themselves at the end of some weeks, and the patient remained well for about six months.

Soon afterwards she suffered from persistent general lassitude, oppression, and inability to indulge in the least exercise, or to follow her occupation without being rapidly fatigued and enervated. At the same time she had dyspeptic troubles, digested badly, with weight in the epigastrium and feelings of nausea ; also some anorexia and constipation. Towards the middle of 1883 she found with dismay that she was still increasing in size, and that her physiognomy had changed to such an extent that she was no longer recognisable. The feet and hands increased in size, and were for six months the subject of severe pains. These pains appeared almost regularly about three o'clock in the night, and persisted to eight or nine in the morning. They woke the patient up and prevented sleep. It felt to her as if an animal was gnawing the hands. The pains were sometimes so severe that she was obliged to cry. The arms were so feeble that all work became very painful; the fingers became clumsy, and unable to do delicate work.

None of these phenomena were manifested in the lower extremities.

During the following years the extremities (head, feet, and hands) have increased to a very considerable extent; she is no longer able to wear the same gloves, and has been obliged to increase the size of her stockings and bonnets.

Her body, and especially her face, is enlarged to such an extent that several people who had not seen her for some years failed to recognise her. The pains in the hands had disappeared for about twelve months, and had been replaced by erratic and often violent pains, which lasted some days, and were situated chiefly in the calf of the leg, the thighs, and the shoulders. She had also very bad dyspeptic troubles, and a feeling of lassitude, more marked at times. There were fugitive attacks of headache and neuralgic pains in the scalp. Sight somewhat feeble; occasional slight fog before the eyes, which rendered objects confused. As regards the ears, there were humming and tinkling noises in the right ear, with diminution of the hearing. These symptoms might be due to a purulent catarrh, which involved both ears two years ago, especially the left.

There were somewhat troublesome palpitations and slight sense of oppression. No trouble with the urinary organs. Since the onset of the malady, occasional very abundant and troublesome sweats. No convulsive seizures nor vertigo; always some giddiness. She has never been nervous, but has, however, for some time been a little irritable and depressed, possibly on account of the different social relations caused by the malady.

Present condition.—Head an enormous size. Forehead large and elevated, frontal eminences thickened; parietal eminences very marked, as well as the outer occipital protuberance. The whole cranium is increased, especially at its sides.[1]

[1] Measurements of cranium given.

On the face the hypertrophy has specially involved the median parts. The cheek-bones are slightly thickened. The eyes and eyelids are normal. The cheeks themselves are flat and not swollen. Conjunctival folds normal. The "nose," on the contrary, is enormous, aquiline, and well developed; nares widely dilated. From the root of the nose to the tip is 10 cm. The lips are thickened, especially the lower. The development of the lower jaw causes the chin to project forwards. When the jaws are closed, the incisors of the lower jaw are found in front, and separated from those in the upper jaw by about 2 cm. The molars alone meet one another. The outer third of the upper jaw can be placed on the inner third of the lower jaw. The gums are equally modified. The teeth are separated a distance of some millimetres, whereas formerly they fitted exactly, and were, according to the patient, very close together.

There is very marked "macroglossia;" the dorsal surface of the tongue is fissured; speech and mastication difficult.

The abnormal size of the hands and feet attracts attention. The hands are not lengthened, but are widened and thickened. The fingers are large and rounded, and this increase in size interferes with most of their movements. The thenar and hypothenar eminences form considerable projections; the palmar folds are very marked, causing the parts they separate to be more prominent. Nails normal. Wrists somewhat large.[1]

The feet are, like the hands, increased in size. The toes are large and thick; the hypertrophy of the great toe is very marked. The whole of the plantar surface of the foot is crossed by more or less deep fissures. The malleoli are large. The tibiæ, tarsal bones, metatarsals, besides the bones of the carpus, metacarpus, and phalanges, are all thickened. The skin of the hands, feet, lower part of forearm and leg, is thickened, less pliable, and somewhat hard. The giant size of the extremities produces an extraordinary appearance.

Her disposition, which was formerly quiet and peaceable, is distinctly more irritable. Height 1 m. 57. Neck thick

[1] Measurements of hands given.

and short. A small goitre the size of a small Tangerine orange occupies the middle line. The lateral lobes have a normal size.

The clavicles, ribs, and different portions of the sternum are not thickened. The lower part of the thorax is enlarged. Nothing particular to note respecting the spinal column. The patient is a little round-shouldered. She continually wears a corset, without which the chest cannot be carried upright; when she is not wearing it she states that her body is bent double. Nipples normal. Abdomen large; its circumference at the umbilicus is 1·25 m.; it is somewhat prominent, so much so as to have suggested pregnancy. The subcutaneous fat seems to account for most of this fulness. The buttocks are large. It is difficult on account of the fat to distinguish any alterations in the pelvis: the iliac crests are more distinct than usual. The labia majora are slightly thickened. The upper limbs share in the general fulness; also the lower limbs. The circumference of the left calf is 30 centimetres. Patellas normal. Tibias only slightly enlarged. On the right thigh there are some varicose veins. No affection of sensation. There is a continual sense of weariness. When the patient works she is quickly tired; the arms cannot sustain effort long. The overgrowth of the hands impedes sewing or knitting, causing her fingers to be clumsy. Occasionally she feels pricking sensations, swelling and heat in them, never any symptoms of paralysis. There are erratic shooting pains, especially in the lower limbs, also frequent lumbar pains. Frequent and sometimes violent cephalalgia. Sight feeble, with other ocular troubles. A slight degree of anæmia. Taste intact. With regard to the ear, there are some buzzing noises. Examination of the ears by Dr. Launois showed the right membrane to be perforated; the long handle much congested, and displacement of the bright spot. The watch is only heard at 1 centimetre. On the left side the membrana tympani is universally dull, also perforated, but less so than the right; the bright spot has completely disappeared, except at the periphery. Some otorrhœa. On the right side there is pain as if an abscess was about to form. The Eustachian tubes are open on both sides. No disturbance of equilibrium.

No alteration in the voice, or defect of speech. Skin intact. On the least exertion there is abundant sweating; also a condition of continual moisture.

Some slight palpitation and attacks of oppression. The patient cannot run or go up stairs rapidly. Pulse regular, 70. Auscultation of the heart shows nothing definite.

In the lungs expiration is slightly prolonged. No disturbance with the urinary organs. No sugar nor albumen in the urine. The intellect of this patient is quite good, and does not appear to have undergone any change. She easily understands what is said to her, and answers intelligently; no other nerve disturbance.

There is no sign of the existence of the thymus.

CASE X (Farge).—Cad— (Louis), 31 years of age, born at Montiers (Loire-Inférieure), was admitted into the Hôtel Dieu, No. 4, Saint André Ward, in February, 1889.

His father was 1 m. 70 in height; he died, when old, from pneumonia; his mother was only 1 m. 40 in height. He has a brother whose height is 1 m. 64. Louis Cad— was small up to the age of seventeen years; he states that he then weighed only 35 kilogrammes. All at once about the age of eighteen, and without appreciable cause or symptoms of any malady, he began to grow very rapidly. His height increased in two years to 1 m. 68. At twenty his weight attained 75 kilogrammes. Till twenty-three he appeared to make no fresh progress, and his weight and height remained the same. He could easily wear his brother's hat, who was smaller than he. He believes, however, that since this time his feet and hands have become larger than normal. His muscular power is great. At twenty-three, while cutting down a tree, he was knocked down by the trunk, which injured his hip and right side. He was confined to bed for ten months, and on getting up it was found that there was considerable curvature of the spine in the neck and back. He states that during the period between twenty-three and twenty-four his head has attained its present dimensions. His appetite has become considerable, and his weight has increased to 96 kilogrammes. During all last year he has complained of dorso-lumbar pains, which were treated by subcutaneous injections of nitrate of silver, with no other result than the abscesses which followed. He has never had any head or joint pains. During the last five years, except for a bad attack of variola, he has been able to continue his work, and has not complained of any inconvenience beyond that caused by the kyphosis. At the time of his admission to my out-patients for bronchitis he complained of general weariness, and specially of a considerable stiffness of the vertebral column, which made it impossible for him to bend down sufficiently to raise anything from the ground. He was more thin and weak than usual. He has never had rheumatism nor syphilis, and presents no traces of them. There is nothing in the physical signs to indicate alcoholism.

Present state.—February, 1889. Height 1 m. 56. His head is enormous, and is placed between high shoulders. The limbs are large and short. The physiognomy, though of a low type, is not deficient in intelligence; the voice is deep and harsh.[1]

The nose is short. The mouth opens excessively wide. Both lips are equally large, the lower more overhanging. The tongue is broad and large, of regular shape. Pronunciation is clear.

Trunk.—The dorso-lumbar curvature extends to the prominence at the base of the sacrum; the kyphosis has no angular projections; no lordosis.[2]

The hair is abundant, very dark and coarse. There are no crests to the sutures. The forehead is not depressed, and the superciliary ridges do not form any considerable projection. The neck is thick and short, but without any hypertrophy of the thyroid. Respiration is only altered by the severe bronchitis, for which C— has been admitted to the hospital. Cardiac functions normal. I have carefully searched for the post-sternal dulness of Erb, and have only found very slight evidence of it, except at the right border of the first part of the sternum. The abdomen is large, but does not project more than the kyphosis. The skin is thickened on the face by variola, and shows a number of black specks; no œdema or discoloration. The muscles are

---

[1] Detailed measurements of head given.
[2] Measurements of chest given.

flabby and the muscular power diminished. No affection of the sight.

The patient lost the hearing of the right ear for a time from an abscess, but it has been nearly normal since. Taste and smell are not affected. I have not found on the skin any anæsthesia or tender spot. The sexual functions are rather feeble. There is no trace of varices.

CASE XI (Flemming, 1).—A woman æt. 45. No hereditary or personal antecedents of importance; no syphilis. In 1881 her menses began to appear in an irregular and intermittent manner. Two years after, at the age of thirty-seven, they ceased completely. At this time, 1883, the patient noticed that her hand began to grow. At this time she suffered pains of a lancing character in her forearm and fingers. Two years after this increase invaded the feet, and finally in 1886 the face began to change in form and dimensions. Since then these three extremities have remained in the same condition. No alteration in the sight has occurred since 1888.

Present condition.—Cephalic extremity.—The face has the form of a long oval. The cheek-bones are prominent, especially the right; the lids are also thickened. The nose is hypertrophied in all its dimensions; its cartilages are thicker and more resistant. The lower lip is very prominent and overhanging. The tongue is broad. The lower jaw is increased in size, massive, and consequently the lower teeth project in front of the upper. The palatine arch is very high. The thyroid gland is normal as far as can be judged.

Upper limbs.—The hands are enormous and flattened; also the fingers, on account of the hypertrophy of the soft parts which form distinct pads. The muscles are well developed. The long bones have preserved their normal length.

Lower limbs.—The feet are enormous, flattened and widened, presenting furrows and pads like those of the hand.

Thorax.—Post-sternal dulness. The clavicles and ribs are thickened; also kyphosis of the upper dorsal region.

Special senses.—Sight: With the left, $\frac{1}{50}$; atrophy of the optic nerve; dilatation of veins. With the right, $\frac{2}{50}$; less advanced atrophy of the optic nerve. Marked retraction of the visual field, especially on the right side. The other senses are normal. The patient perspires copiously. The quantity of urine is normal; no albumen; no sugar; urea of normal quantity. The temperature is subnormal.

CASE XII (Verstraeten).—On August 14th, 1888, I was surprised to see among my out-patients a young woman with a very large and swollen face, and enormous hands and feet. Her face was covered over with well-developed hair. Her gait was like that of a man; and her voice deep, with no feminine tone to it. This was sufficient to give probability to the diagnosis that it was either a case of myxœdema or of acromegaly. The woman complained of a violent obstinate headache, which had resisted a variety of treatment, and which was the cause of her visit. As regards the extraordinary development of the nose, lips, hands, and feet, it was necessary to mention these before she gave any attention to them. She was not, however, lacking in intelligence or discernment, but her violent headache made her forget everything else. The following is the account I obtained from her:—

Mdlle. L—, æt. 29, single, is a tailoress by profession. Her menstruation had become suppressed following a severe chill, and since then she had suffered intensely from her head. There had been nothing particular to note in her history before her malady. Her mother stated that she could walk by herself at the age of three months, like all her brothers and sisters. She was a precocious infant, big for her age; she was thin, and had always had a somewhat tall figure. Her menstruation commenced at the age of sixteen; since then the periods have appeared at intervals of five weeks and a half, and lasted five days. She does not lose much blood, not so much as her sisters; never leucorrhœa. Both parents are living: her father, aged sixty-seven, of medium height, has been suffering from general anasarca for the last three months; he has taken alcohol largely. Formerly he was strong and of good constitution. He died January, 1889. The mother, aged fifty-nine years, a little above medium height, has lost all her teeth, has a goitre on the right side, and is subject to have the fingers go white under the influence of cold (vascular spasm). Mdlle. L— is the fifth of a family of twelve children, of whom four died young. Her three brothers are in good health, and taller than she is. Three sisters are also in good health and well grown. One sister of twenty-two is tall also, but is subject to frequent attacks of simple purpura.

As regards uncles and aunts, both paternal and maternal, they appear to have somewhat feeble health, and dropsy seems to be a frequent malady in the family. No one else has suffered from any illness like this patient. The latter resembles physically her mother rather than her father.

On January 23rd, 1886, it snowed and was extremely cold. Her sister was married on the same day, and she became greatly fatigued and suffered much from the cold. Her feet were literally frozen, and she burnt them badly in attempting to warm them.

She lived besides in a damp and cold house, worked hard from five or six in the morning to eleven in the evening, and went out but little. Her last menstrual period had been January 10th, 1886. Since then her menses had only appeared on April 7th of the same year, but were less abundant. Mdlle. L— suffered from a series of depressing emotions, which greatly upset her; they were specially caused by the alcoholic excesses of her father.

On April 10th, 1887, her menstruation occurred for the last time; she lost only a little blood, very dark in colour, at this time. She noticed sometimes some drops of blood when blowing her nose, which had also appeared formerly with rheumatism of the neck. She had had no other supplementary hæmorrhages.

In October, 1887, she was persuaded to take large quantities of milk and beer, as a sure means of bringing back her menstruation, but without any benefit.

The headache is occipital, radiating towards the eyes; when she lowers her head the pain increases. She refers the deep twitchings to the occiput and upper part of neck. She suffers besides from a weight rather than a pain under the orbit; work, confined atmosphere, and heat increase the pains. She feels relieved when she is in a wind, and when her mind is rested. During night she sleeps as well as ever; but her nights are short, all her time being occupied in work. These headaches have lasted more than two years, and are specially bad when she gets up. She has never suffered from migraine. For two years also she has been subject to palpitations of the heart, which return from time to time, but are specially bad on going to bed. "I feel then as if something gripped me over the region of the heart, and prevented me breathing; fortunately this does not last long, sometimes a minute only, sometimes half an hour or more." These attacks seem to be increasing; their onset occurs very irregularly, and at intervals of longer or shorter duration, but without any appreciable cause. The intervals may last one month, fifteen days, eight days, or only twenty-four hours. She complains also of pains situated in the back, opposite the lumbar region, and radiating to the side of the right iliac fossa. They date their commencement from the same time as the suppression of the menses; they increase when she walks much. Sometimes she experiences pain in the legs and at the soles of the feet, especially on the right side, and independently of fatigue or wearing tight shoes. She has always had the masculine appearance which is observed now. The hairs which disfigure her face are certainly small and scattered, but for two years they have much increased. The face besides is totally changed. A portrait taken in March, 1883, represents her comparatively thin and meagre, with small nose, normal hands, and long flexible fingers. She was then twenty-four years old. This portrait, according to her account, is a good representation of what she was then.

A second portrait, dated October, 1886, shows her already considerably grown; but the nose, ears, chin, lips, and height present nothing abnormal; in fact, it is the portrait of a well-made person.

In September, 1888, at my request, Mdlle. L— again went to the photographer who had taken the second portrait. This shows well the principal features of the malady at that time: the large ears, enlargement of the lower part of face, thickened nose, very thick lips, massive hands traversed by furrows, and very considerable height. Her weight was then 74 kilogrammes.

A fourth portrait was taken at the end of December, 1888. The disease is still more advanced. It has lost the characters which might suggest myxœdema, retaining those only of acromegaly, which are very distinct; *i. e.* the lengthened face, big nose, swollen lips, much lengthened lower jaw, enormous hands. The weight of the body had diminished to 68 kilogrammes. Formerly she ate but little; for two years she has been insatiable, so that it is a real torment to her; she has a preference for bread. Thirst is not increased. The motions are normal. The urine is more abundant since the commencement of her malady, but micturition is not more frequent than usual. She perspires as much in the day as the night. The sweat has no special odour or colour.

The fingers frequently lose their colour; it is sufficient for her to keep them still for a time for them to become as pale as wax. Even before her malady began she was subject to this inconvenience, but it occurred less often. When the temperature is very cold the ends of the fingers become like wax, numb, and deprived of all sense of touch.

The hands are habitually hot and dry; they perspire very little. They swell during walking, the right particularly, and more especially in summer when it is warm.

Excepting at these times, when the hand is numb or very swollen, it preserves all its dexterity in handling a needle and for other manipulations.

The feet have become also larger; the shoes which she wore at the wedding of her sister were enlarged in January, 1887. During the summer of 1886 she has experienced pain, particularly in the ends of her feet near to the great toes. She has, however, continued to walk in these boots during the whole summer. She attributes this pain to swelling of the feet on account of the head (summer, 1886). The bootmaker states that the foot of Mdlle. L— is much increased, and especially in thickness. Two years ago the feet measured 40 points, now they measure 42; the increase in width is relatively more important.

She does not know if the proportions of the neck are much changed, but she fancies so. Four years ago she had a painful swelling at the right angle of the jaw, which she attributes to caries of the molar teeth. These glands have now disappeared without leaving a trace; but there is a goitre on the same side, which her sister noticed two years ago. It is to be noted that syphilis may be definitely excluded.

The width of the shoulders, she states, has not changed. The breasts are distinctly smaller, notwithstanding her general growth.

Her height has increased in extent 5 centimetres. She insists strongly on her hips having enlarged, which is due to considerable overgrowth of the muscles. The hypertrophy involves not only the soft structures, but also more particularly the bony skeleton.

Her height has increased also of late years. She estimates the increase at five centimetres.[1] The shape of the hands differs slightly; the fingers of the left hand are thinner and longer. The palm of her hand is hot and red; the thermometer placed in the right hand for a quarter of an hour registered 36·35°; on another occasion 36·80°. Placed under the armpit during the same time, the same thermometer registered 37·05°.

The tactile sensibility is normal.

The skin forms very thick pads over the palmar surface of the heads of the metacarpal bones; it is not callous, but pliant and red; its thickness seems increased. The thenar and hypothenar eminences are well developed.

The nails are overlapped by the soft structures; those of both the thumbs are flattened, transversely striated, and roughened; they appear short. Those of the middle and index finger are also flattened, but not striated, smooth, and slightly enlarged. Those of the ring and little finger are but little modified beyond being smooth and somewhat curved.

The changes in the nails are much less marked on the left hand than on the right, on which the other symptoms of acromegaly are also much more marked.

The skin of the fingers is thickened, dense, and firmly adheres to the ungual phalanges; there is no œdema. Its colour is red on the palmar surfaces and sides; it is pale on the dorsal surface, and specially over the joints.

[1] Details of measurements of the body given.

The general form of the fingers is regular; the soft parts are swollen up to the first phalanges on the palmar surface. This swelling is less marked on the second phalanges. The left ring and middle finger have enormous ungual phalanges. This feature is less marked on the right side, on which, on the contrary, the index finger is very large at its extremity.

The bony structures manifestly share in the hypertrophy which involves all the tissues. The projections at the ends of the bones are well developed. This development is also very marked on the styloid process, olecranon, epicondyle, and trochlea. The patient asserts, however, that she has always had these bony outgrowths, which have not increased in size since her illness. The arms appear thin and weak compared to the hand; the muscles are certainly not hypertrophied; they are somewhat diminished in size, according to the patient.

The thorax is large and well developed. Respiration is of the costal type; the lower part of the thorax hardly moves at all. Resp. 20 per min. The clavicles are strong; their curves are not increased; they measure with the compass D. 150 mm. and G. 145; the sternum is 17 centimetres long excluding the xiphoid appendix, and strongly curved. It presents five transverse ridges on its anterior surface, which are prominent to the touch and placed 3 or 4 centimetres apart. The first and fourth of these crests are the most definite; the manubrium sterni is very large, and seems hypertrophied. The body of the sternum appears of normal size. The xiphoid appendix is very elastic and flexible.

The costal cartilages present nothing special except those of the twelfth ribs, which are distinctly hypertrophied. There is nothing to note concerning the ribs, whose shape and form seem normal; nor of the vertebral column, which is also normal.

The skin of the thorax is brown and much pigmented; it is also covered with numerous flat warts, of which some are black and others dark brown. They vary in size, some being more than a centimetre across. They are specially numerous at the base of the neck and the top of the waist, but they are found over the whole thorax. They are not of recent development. The breasts are small and soft; their glandular portion is distinctly atrophied. The nipple is well developed, possibly hypertrophied; it is certainly larger than normal, and is surrounded by a number of strong hairs of dark colour. The skin of the chest has no very thick layer of fat. The axillary secretion is sometimes acid and sometimes alkaline.

Percussion is normal over the lung except at the upper part of the sternum, where the note is dull. This dulness is like that described by Erb. Auscultation over this dull area reveals no special bruit; the cardiac murmur is not heard here, or only at a distance. The heart at other parts shows no abnormality. No substernal pulsation. The cardiac impulse is increased, and the precardial dulness is much extended. There is no increased overlapping of the heart by the lung. The vesicular murmur is very feeble.

Examination of the abdomen reveals nothing special beyond slight development of the adipose tissue. The abdominal viscera seem normal on palpation.

Vaginal examination shows considerable development of the clitoris, which appears three times its normal size. The labia majora are not hypertrophied. The hymen is intact, but its aperture is sufficiently dilatable to allow of exploration without rupture. The uterus is raised; its neck is small and soft, and is found towards the right iliac fossa. The body of the uterus seems small; it is found in the right iliac fossa. As far as can be judged the pelvic bones are thickened, but the cavity of the pelvis is not large. The subpubic arch is somewhat rounded, and the lower aperture appears narrowed.

Mdlle. D— states that sexual desire has never been very strong with her. She refused an offer of marriage because the menses were suppressed. The urethra is much developed, and distinctly hypertrophied; on palpation it appears as large as the little finger. It is soft and tortuous. Nothing abnormal can be detected with the bladder. There is no trouble in micturition. The urine has been examined by different methods several times during the malady. Its quantity, measured during several consecutive days, varied from 1500 to 2000 cubic centimetres. It is clear, of faintly yellow colour, without deposit, forming a slight cloud on standing;

feebly acid or sometimes neutral; sp. gr. 1012. Urea in medium quantity, 13·33 gr. Sometimes a small quantity of albumen; never any sugar. With the microscope numerous cells are found, with definite nuclei and somewhat granular, some having the shape of a club; also a few blood-corpuscles.

The lower limbs are specially modified at their extreme ends. The legs as far as the knee appear normal; it is possible that the condyles of the femur are a little increased in size. As regards the great trochanters, I believe they have undergone still less alteration, and the patellæ have also preserved their proper shape and size.[1]

The nails are wide and short; those of the great and second toes are transversely striated. The general hypertrophy of the foot is evident at the first glance. It reaches its maximum at the great toes, of which the dimensions are enormous. The projections at the heads of the first metatarsals are also increased. The sole of the foot is muscular; it is to a certain extent flat-footed, with thick folds of skin. The latter is much less thickened on the backs of the feet, where it seems normal; also on the first phalanges of the toes. On the other phalanges the skin is more dense, and is with difficulty raised from the deeper structures. At no part is there any sign of œdema. The skin of the leg is not altered, but the cutaneous veins are numerous and dilated, and the subcutaneous fat is well developed.

The temperature of the foot does not seem to be increased on touching it; no examination has been made with the thermometer. "Whilst the hands are habitually warm and dry, the feet are rarely hot, and then they perspire."

The patellar reflexes have been frequently tested; they respond very feebly, being slightly present on the right, and almost absent on the left side. The plantar reflex is very feeble on both sides; it is the same with the reflex of the tendo Achillis.

Electrical excitability is decidedly diminished, but the other electrical reactions are not altered.

Examination of the head shows plainly the nature of the malady. The cranium is of somewhat large dimensions, but, according to the patient, has not increased during the last

---

[1] Measurements of the leg given.—TR.

two years. It is covered with thick hair, of a brownish chestnut tint, which has not changed during the malady. The sutures are normal. The occipito-parietal alone forms a ridge appreciable to the touch. The occipital protuberance alone forms a very distinct horizontal crest. Mdlle. L— cannot state whether these phenomena are of recent date, her attention not having been drawn to them. The face has an oval form, with large upper end. When she first came it had a more rounded form.

The forehead is fine and elevated, but already marked with wrinkles. The orbital borders and superciliary ridges do not project abnormally.

The latter are hidden under thick eyelashes, which join in the median line. The upper lids are swollen, especially in the morning; or rather the skin above the lid forms a fold over the eye. The lower lid is not altered. The eyelashes are well developed; the eyes are greyish blue, and the pupils react well to light. The nose is thick, short, and flat. The cheeks are a little prominent. The mouth is large and the lips thickened, especially the lower. The chin is rather prominent. The submental fold is very marked, so that the chin is almost double. The ears are well formed, possibly a little enlarged. The skin of the face is hot, pale, of a dull white colour, and slightly dusky. The skin appears very thick; the subcutaneous fat is well developed under the chin.[1]

The teeth are white, small, and regular; they are well preserved both in front and laterally. Most of the molars are carious, including all those in the lower jaw. The two jaws are prominent; the lower overlaps the upper at least 2 mm. The tongue is thickened in the form of a club. No special markings on it beyond some deep fissures at its middle; it is also very long. The patient is sure that the increase of her tongue only dates since her malady began. The uvula is also large; the palatine pillars are normal. The tonsils are not swollen. The reaction of the parotid saliva is distinctly acid, that of the sublingual slightly so.[2] The sight is feeble, which the patient attributes to working in the evening. She is certain that prolonged work fatigues

---

[1] Measurements of face given.—Tr.
[2] Measurements of the tongue given.—Tr.

the sight, and believes the eyes have suffered for two years. Examination of the eye has been made by the distinguished specialist, Dr. Van Duyse, and nothing abnormal was found. Hearing, taste, and smell are normal. The circumference of the neck is 37·30 cm. There are no cicatrices or enlarged glands; the skin is very brown. The larynx is hypertrophied, and of a masculine type; the pomum Adami is as prominent as in a man; the hyoid bone does not appear to be attacked by the malady. On the right of the larynx is a goitre, the size of a hen's egg; it is easily seen and felt. The sterno-cleido-mastoids are somewhat marked. The cervical spine is apparently not changed.

For two years Mdlle. L——'s voice has become deeper, and slightly thicker.

Her height equals the measurement between her extended hands.

CASE XIII (Virchow).—Westph., a wrestler, well developed during infancy. Genital functions normal, neither in excess nor diminished; six children, of which none are the subjects of acromegaly. "Muscular power enormous." He can carry easily on his back a weight of 400 kilogrammes. Head large; circumference 655 mm. Antero-posterior diameter of the head 229 mm.; transverse 168 mm. Cephalic index 73·3 mm. Height of body 1838 mm. Hands and feet very thick and large. Fingers and toes enormous. Circumference of the phalanx of the thumb 100 mm. Circumference of the phalanx of the index finger 114 mm. Nose, lips, and lower part of face much hypertrophied. No dulness at upper part of thorax. Thyroid gland small.

CASE XIV (Freund).—On March 25th, 1872, a woman came to the author's consulting room, of medium height, strongly bent forwards, walking awkwardly, and whose head was extraordinarily large and malformed. Her hat, which looked at first like that belonging to a child, was, however, similar to those then worn. Her huge hands, crossed below the waist, were only half covered by her unbuttoned gloves. Her clothes seemed nowhere to fit.

M. Freund almost fancied he had before him a kind of

anthropoid ape, clothed in human dress and walking like a man.

This woman, æt. 34, stated that her parents were well formed, as well as her brothers and sisters. Till the age of seven she was somewhat small. After this date she grew very rapidly. Her second dentition took place at the right time. During the following years (exact dates not known) the size of the hands and feet was remarked upon, but she kept perfectly well. The face at this time showed nothing abnormal. About the age of fourteen menstruation commenced; at first the periods appeared in advance, later on they became retarded. They lasted four or five days, with discharge of varying amount and without pain. At fifteen years she had suppression of the menses for some months; she then noticed an increase in the size of the feet, hands, and face, which was at first attributed to her general growth and to adipose tissue. The menses then appeared more and more irregularly; they were absent for intervals of six months or more, and finally completely disappeared at the age of twenty. She was married, at twenty-two, to a man who was then in good health; she had no children, and later became indifferent to sexual intercourse. The muscles of the hands, feet, and face then became more and more enlarged. Then also she had difficulties at the dressmaker's, milliner's, shoemaker's, and glover's. Little by little the head became almost too heavy to carry, so that she could no longer hold it upright as formerly. Manual labour soon became impossible, because the hands could neither direct nor hold the needle, and because the weight of the hands caused them to be easily fatigued. Walking became awkward and without elasticity. An almost constant sensation of heaviness and exhaustion was experienced, sometimes changing to severe lancing pains in the neck and limbs, so that instead of being lively and active as formerly, she became morose and melancholy. This was increased by her deformities, and she became depressed and misanthropic, shunning the sight of other people. Examination gave the following results:—She was a fair-complexioned woman, of vigorous constitution. The small size of her arms and legs is striking, which causes the difference in size between the different segments of her limbs to be prominent. On the contrary, the soft structures of her hands and feet have undergone a great development, so that the skeleton of these parts is only examined with difficulty. The skin is specially soft and elastic, not œdematous, and on the hands and feet thrown into huge folds. The colour of the skin is that ordinary to pale blondes. There are no visible varices. The mucous membranes are pale. The expression of the face is sad and nervous. Speech is somewhat embarrassed. The organs of sense have normal functions. The mental faculties are not affected. Sleep generally good. Appetite fairly good. Slight constipation. The tongue presents nothing abnormal. The deformed appearance of the body is principally due to the arched curvature of the upper part of the dorsal spine and the special attitude caused by this kyphosis, notably in the pelvis and lower limbs. It is also due to the shoulders being depressed forwards and downwards.

Another striking point, especially when seen in profile, is the disproportionate and almost gigantic development of the four extremities. The entire face, by its huge proportions, causes the cranium proper to appear too small. It is carried strongly forwards and downwards. The hair of the skin, of normal thickness, seems to be more limited in distribution than normal. It is specially the lower jaw which is prominent in all directions, and also the region of the temples and ears. That part where the upper jaw approaches the nose is equally abnormally developed, and gives the impression of having been drawn out from its normal position. The lower lip is a little larger than the upper. The lower dental arch projects in front of the upper about five millimetres. The arms are dependent; the extremity of the fingers reaches twenty-five centimetres below the outer condyle. The first glance shows that the hand and forearm are of disproportionate length, and that the hand is besides excessively large and massive. The clavicle and scapula are also of enormous size. As regards the legs, the great development of the feet is at once noticeable, which have the appearance of flat feet. When the patient is standing the skin at the borders of the feet assumes under the pressure the form of a large pad. The legs, which are certainly too long, are also thin, and the soft structures are flaccid. The pelvis appears larger than normal. The mons veneris forms a well-developed projection. The buttocks are soft and pendulous. The lumbar region of the spine is abnormally flattened; the sacral region tends to assume a vertical direction.

In the neck a great development of the sterno-mastoids is to be noticed, and also of the trapezius in the nape of the neck. The thyroid body presents no appreciable change. The upper part of the thorax is depressed; the lower part is enlarged. The breasts are soft and seem normal. There is nothing to note abnormal with regard to the abdomen. The external genitals seem completely developed. The clitoris appears certainly too large, and presents a thickened prepuce. Around the corona of the clitoris adhere numerous masses of half-dried sebum. The labia minora have a very thick rugose mucous membrane of a brownish-yellow colour. The remarkably large vagina presents but few folds, and terminates posteriorly in a large cul-de-sac. The uterus, in its normal situation, is 7·5 centimetres long, and presents besides all the characters of commencing senile atrophy. The softness and thinness of its walls are specially noticeable. The ovaries, easily examined by palpation, are flat, firm, and their surface somewhat irregular.[1]

Such was the condition of this woman at sixteen years of age. Recently Dr. W. A. Freund has requested his brother, Dr. M. B. Freund, at Breslau, to make a further examination of this patient. Below are the notes on September, 1888. This woman is now aged fifty, but appears much older. She has become withered and senile, and the monstrosity of the face has become still more increased. She has a very marked and uniformly arched kyphosis, which increases still more the appearance of decrepitude. For eight years she has suffered from increasing neuralgic pains, always in the feet, and causing difficulty in walking. This condition is simply a symptom of well-developed tabes. All the reflexes are abolished; and standing upright, and walking with the eyes shut, have become impossible. The walk is typically ataxic, none of the features being absent. The etiology is very doubtful. This woman's husband died eight years ago from a cerebral

[1] Measurements of the body given.—Tr.

affection. She knows that some years after her marriage he had an attack of syphilis, and that she also contracted it from him. She had a rash, and was treated by mercury and iodide; the ulnar gland is, moreover, still indurated and increased in size.

The patient believes that the hands and feet have not progressed for some years, for the size of her gloves and shoes have remained the same. The lower jaw has somewhat increased, for she has been obliged this spring to change her set of teeth, which three years ago fitted as well for the upper jaw as for the lower. I feel sure that not only is the lower jaw larger, but also the whole of the upper jaw, and also all the temporal bone. The prominent border of the frontal bone is displaced abnormally far forward, so that the space between it and the anterior border of the tragus is very considerable. There is considerable thickening all around the ear, and the mastoid region presents considerable curvature, so that the mastoid process itself is placed nearly on the same plane as that portion of the temporal situated behind it. All the temporal region gives the impression of being abnormally large.[1]

On inspecting the mouth the narrowing and very marked arching of the palate is noticed. (The increase in the size of the limbs having, according to the patient, commenced during infancy, M. Freund considered that the syphilis contracted later could hardly have acted as a cause of the acromegaly.)

The mental functions seem intact. The woman has, however, recently made attempts at suicide, on account of pains in the legs (tabes). This woman's mother is still living, and of good build. Her father was also the same; and in the family no malformation has ever been found. The patient would not allow any photographs to be taken at the last examination.

CASE XV (Roth, 1).—M. Br—, æt. 37, a Jewish merchant born at Motilew, was admitted (to the Catherine Hospital) on March 21st, 1884. He was a man of good constitution, very pale, " excessively apathetic " and feeble; he passed most of

---

[1] Measurements of the different parts of the body given.—TR.

his day in bed. He showed, however, no organic lesion, nor paresis or atrophy of muscles. The region of the manubrium sterni (thymus gland) has not been examined in detail. The hands, feet, and head have increased in size. The shape of other parts of the body, especially that of the forearm and leg, is normal. The skin and subcutaneous tissue show no pathological change. The patient has grown little by little for the last three or four years, but remained stationary for the four months preceding his entrance to the hospital. Until 1877, Br—, who was then twenty-two years old, felt perfectly well; the size of the head and the extremities were normal. The onset of the acromegaly coincided with a condition of weakness, so considerable that the doctors attributed it to ague, on account of its more or less intermittent character. Since then this weakness has not left the patient; it is, however, less marked. The antecedents of Br— show nothing as regards etiology. There is nothing of an "hereditary" nature.[1]

CASE XVI (Strumpnell, 1).—This case occurred in a woman who presented all the symptoms known to characterise acromegaly. The case merits special attention from the fact that the patient suffered from "diabetes," and also on account of the profuse perspirations which took place from the whole body. There existed besides a number of subjective symptoms, "cephalalgia;" pain in the neck and back; moral depression, and marked feebleness; disturbances of the cutaneous sensibility, especially analgesia, and diminution of taste and smell. These symptoms speak in favour of the nerve origin of acromegaly.

CASE XVII (Schultze, 2, *résumé*).—A. L—, æt. 33. This man presents most of the changes and symptoms of acromegaly. The cranium presents some deformities. The crest and external occipital line are well developed, so that they form with one another a kind of cross; which was remarked on by the hairdresser of this patient. The face is lengthened, especially its lower part. The nose is thickened and enlarged; it is almost as long as the ears, which are also increased in

---

[1] Measurements of body given.—TR.

size. The lips are large, the upper much less than the lower, which tends to turn downwards. The tongue is hypertrophied; the lower jaw is thickened and increased to a marked extent, resulting in prognathism. The orbital ridges and cheek-bones are prominent. Myopia; temporal "hemianopsia," "atrophy of the right optic nerve;" veins and arteries at the fundus of the eye smaller than normal. Ocular reflexes normal. Neck large and short. Thyroid gland examined with difficulty.

Thorax thickened, broad, and deep. No hump. Hands enormous, enlarged, and thickened; they are also longer than a normal hand (length between the fold of the wrist and extremities of the middle finger is 21 cm.). Marked hypertrophy of the thenar and hypothenar eminences, and of the soft structures corresponding to the metacarpals and fingers; resulting in palmar fissures and articular folds. The fingers are enlarged in such a manner that they are nearly as large at their bases as at their tips. The nails are small in comparison to such large fingers. Wrists somewhat large.

Feet.—The same peculiarities as regards thickening, enlargement, and absence of increased length in their dimensions. Joints normal. Muscles normal, with strength in proportion to their development; movements not altered. Cutaneous sensibility normal. No sugar nor albumen in the urine; no polyuria. Mental and sexual functions intact. Skin a yellow colour; a certain number of tubercles of molluscum fibrosum. More or less ill-defined pains. No paralysis nor convulsions.[1]

Case XVIII (Schultze? *résumé*).—It occurred in a man, æt. 63, who presented on the ends of the limbs (hands and feet) the alterations of "acromegaly." The cephalic extremity was the seat of somewhat marked changes. This man presents a chronic rheumatic arthritis of the wrist and some other joints, but M. Schultze notes that the extremities of his patient were found to be hypertrophied a long time before the appearance of these rheumatic complications. A

---

[1] Detailed measurements of the body given.—Tr.

very curious fact is that two of his grandsons, his daughter's children, also present the changes of acromegaly (? ?).¹

CASE XIX (Adler, 1).—Mme. Anna H—, æt. 34 years, was admitted in the month of January, 1888, into the German Hospital. As in most similar cases the history was deficient. No hereditary antecedents. The father of the patient is still living; he is a man aged fifty-four, very robust, and presenting no abnormality. Her three brothers and sisters are all in robust health. Her mother is dead; it is not known what was the cause of death, but the malady was of short duration.

The patient has never had any serious malady. The menses appeared at fifteen; they were somewhat abundant, very irregular, being often absent for three months. She was married before fourteen; she has never had a child. At eighteen her menstruation ceased completely.

At twenty the legs began to swell; wearing a somewhat tight shoe, however, was sufficient to cause this swelling to disappear. About this time swollen glands appeared over the parotid region on both sides. A little later, often in the morning, the lids became swollen. All these symptoms have appeared and disappeared during a short period.

Concerning the onset of the abnormal growth of the bones the patient knows nothing definite. The progress of the changes was so slow and progressive that the first signs passed unnoticed. The parents of the patient assert that before her eighteenth or nineteenth year she was lively and robust, and ailed nothing.

Between eighteen and twenty, at a date she could not exactly fix, a ring, which she had always worn on the finger without inconvenience, began to press on the skin so that she was obliged to cut it. She was obliged to cease wearing it on this account. It was through this that her attention was drawn to the development of her fingers. About this time she noticed also that the different joints, such as those of the knee and foot, became swollen. It was only a few years ago, it appears, that the patient noticed that the lower row of teeth had advanced in front of the upper.

About five years ago the patient took a long drive in a

¹ Measurements of the body given.—TR.

carriage, the shaking of which caused such distressing pain in the back that she was unable to return home. Since this occasion the patient has suffered from pains in the back and limbs which prevent her walking or standing, and which do not completely leave her even in bed. She is in such a state that the least excitement causes painful crises.

Present state.—The abnormal shape of the head and the enormous proportions of the hands (like talons) at once attract attention on examination. The neck is short and somewhat thick.

The hairs are black, stiff, and well developed. The forehead is low and retreating. The whole of the hairy skin is thickened. One remarkable point is that this hypertrophy corresponds to the occipital bone; the occipital protuberance also presents a very great development. As a result the cranium has the shape of an oval flattened at the sides, with longest antero-posterior diameter.

As regards the bones of the face, one is struck by the enormous development of the lower jaw. This bone is hypertrophied in all its proportions. The lower row of teeth projects forwards, and overlaps the upper, causing considerable inconvenience in mastication. The chin is specially developed. The superciliary ridges and upper jaw are hypertrophied, but to a less extent than the lower jaw. The zygomatic arches are very strong and well developed. The nose is long and thick.

The teeth are separated from one another, and are very carious. The enamel, especially that of the incisors, is worn away. The tongue is broad and thick; its papillæ are hypertrophied to such an extent that it appears on the dorsal surface to be traversed by ridges and furrows.

The mucous membrane of the lips and mouth is pale. The pallor and softness of the skin of the face is striking; on examining it closer it is found to be thickened, especially near the lower lids and zygomatic arches. In fact, the face presents a peculiar œdematous appearance, although at no part does true œdema exist.

The parotid glands appear hypertrophied and enlarged; they are not, however, painful. The submaxillary and cervical lymphatic glands are numerous and large; the thyroid gland is slightly developed. The latter is easily examined in all its relations, but is certainly a little more developed than normal, such as is frequently the case in women in good health.

On examining the patient in an upright position a marked kypho-scoliosis is seen, resulting in deformity of the spine. The ribs are throughout noticeably thickened and widened, and it is very probable that they are also increased in their longitudinal diameter.

The clavicles are thickened, especially at their sternal extremities. The mammary glands are atrophied, though the skin covering them is thickened. Over the whole trunk the skin is remarkably pale and thickened; the adipose tissue is moderately developed. The bones of the pelvis are thick, and the iliac crests extraordinarily large. The whole pelvis presents the scoliotic type.

As regards the lower limbs, what at once attracts attention is the huge proportion of the feet. Their entire skeleton is remarkably enlarged and thickened; the skin is equally so. As a result of this the feet are enormously long and wide, and more arched than normal. The nails of the toes are large and smooth. In comparison with the feet and the rest of the body the legs appear somewhat small. The muscles of the leg are very atrophied and weak. The bones are much thickened, especially near the epiphysial extremities. The patellas are thickened and enlarged. The thickening of the skin over them is less than on the legs.

Over the whole body slight pressure on the bones causes violent pain. This pain is most pronounced over the hands, feet, and in the neighbourhood of the joints and sternum.[1]

Examination of the viscera shows nothing peculiar. The lungs are healthy. The cardiac bruits are normal but weak. The pulse is regular but small; it beats 76 to 80 per minute. The various mucous membranes are pale; the patient is somewhat anæmic. The temperature is normal. The dulness which is found in cases of persistence of the thymus, and which Erb has noticed in his patients, is not in the least degree present in this case. The liver, the spleen, and the kidneys are normal. The urine, which has a specific gravity of 1018 to 1022, contains no albumen. Repeated chemical

[1] Detailed measurements of the body here given.—Tr.

and microscopic examinations of the urine for a long time have failed to detect any certain abnormality in the urine. On the trunk and extremities a number of lymphatic glands are found, increased in size and hard, but not painful. Ophthalmoscopic examination gives normal results. The skin and tendinous reflexes are normal. No contractions.[1]

As regards the intellectual faculties the patient shows various symptoms of mental weakness. Whereas formerly she was lively, and showed somewhat animated spirits, she is now apathetic and dull. She takes little or no interest in what is passing around her, and speaks very rarely. She produces the impression of stupidity. She is contented, acknowledges the least attention, and never complains of her condition. Her mental state recalls that of cachexia pachydermia and cachexia strumipriva.

The subjective phenomena are chiefly a general weakness and pains resulting from the weight of the body. She cannot stand upright nor walk without assistance. She passes her life half reclining on a bed. She can sew and do crochet for a very short time. She passes many hours of the day with her eyes fixed vacantly before her. Once or twice a week she has violent headache, which is localised specially at the back of the head, and which is relieved by antipyrin. She suffers besides from habitual and obstinate constipation, which has to be continually combated with all sorts of purgatives. Finally, it may be noted that during the first six months of her stay in the hospital a slight increase in the size of the hypertrophied parts was noticed. For some time there seems to have been a period of "arrest in the progress." Observation will show us whether this improvement will last or be only temporary.

CASE XX (Erb).—A woman æt. 58; no hereditary antecedents; no analogous affections of her children or relations. The onset was at the age of forty-eight years. From forty to forty-eight she suffered from frequent attacks of migraine.

With the appearance of the menopause at the age of forty-eight an increase in the size of the face and hands occurred,

[1] Electrical reactions given, showing decreased reaction to the faradic and galvanic currents.—TR.

and at the same time also that of the feet and trunk. In from three to six months this increase attained a marked degree, then in the three or four following years its progress was slow. For the last six years the growth has ceased, and also for a year the lower lip has lost a little in size and thickness.

This increase in size was accompanied by itching and numbness. There is overgrowth and projection forwards of the lower jaw, so that the teeth can no longer approximate. The tongue has become thicker, the voice has a lower tone, the face is more brown, and the head is also larger.

The hands and feet appear blue, cold, thickened, and shapeless. The nails are wider, but not longer than normal; the hands are clumsy.

For four years the pains in the arms and legs have disappeared. Since the onset of the malady there has been general weakness, sleepiness, want of memory, inability to think, and headache between four and eight in the morning. Vision is not good, being somewhat dim.

There is great sensitiveness to cold. Perspiration abundant.

No former injury, nothing indicating the existence of syphilis.

Present state.—Height 152 centimetres, weight 87 kilogrammes. Speech a little slow. The cerebral functions are not notably affected. The head is increased; the face is long and broad, but its general form is somewhat oval. The nose is very broad and thick. The lower lip is very thick and slightly pendulous. The skin of the face is thickened, and presents a yellowish-brown coloration. There are deep wrinkles on the forehead; the lids are thicker than usual. No marked development of the subcutaneous connective tissue. No trace of œdema. Skin soft. Ears normal in dimensions and colour. Hair and eyelashes well developed. The malar region forms a distinct projection, the malar bones are more massive and longer than normal. No very marked alteration in the upper jaw; most of the teeth are deficient, and the alveolar processes are atrophied.

The lower jaw is lengthened and considerably thickened, especially in its lower two thirds. The row of teeth in this jaw is about a centimetre in advance of that in the upper jaw.

The width of the nose at the nares is 5 centimetres. The tongue is very large, causing the floor of the mouth to project when the mouth is closed. The palate and the uvula appear distinctly enlarged. The latter is at least double the ordinary width. The tonsils are large and indented.

The larynx is somewhat large. The thyroid body seems completely wanting.

There is very marked thickening and enlargement of the inner ends of the clavicles. The sternal portions of the ribs are also distinctly enlarged. The nape of the neck and the back are very large; the latter is kyphotic in the dorsal region. The thorax is large, broad, and fixed; its circumference about 100 centimetres. Nipples soft and pendulous.

The acromion is perhaps thickened; the humerus does not appear to be so at any part. The bones of the forearm are enlarged and thickened, especially towards their peripheral extremity. The same also in those of the hand and fingers. The soft structures are also hypertrophied; in the hands this involves specially the subcutaneous tissue; in the forearms—the muscles.

The fingers are much deformed, especially opposite the last phalanges; resulting principally from the thickening of the pulps. The furrows between the last and middle phalanx have almost disappeared. The nails are all longitudinally striated, very short, and broad. The muscles of the shoulder and arm are in good condition. The flexors and extensors of the forearm, the size of whose muscles is increased, appear on the contrary somewhat weak. Dynamometer with the right 12°, with the left 11° (compared to 30°—50°, the normal number). The small muscles of the hand perform all their movements; though somewhat weak, they are of normal size.

The mechanical excitability of the muscles is normal; the tendinous reflex of the triceps is present.

There are slight disturbances of sensation, especially on touching the forearm, hands, and ungual phalanges.[1]

The skin of the abdomen seems also a little thickened.

[1] Measurements of the body given.—Tr.

The lower limbs present a distinct increase in size in the legs and feet. As regards the thighs, it cannot be asserted that they are hypertrophied.

There are a number of varices on the legs. The colour of the skin is normal, a little bluish on the feet. The bones are certainly hypertrophied to a marked degree, especially in their lower thirds and at their peripheral extremities; without, however, presenting any increase in length. The tibia in particular is distinctly thickened and enlarged in its lower third; its border is a little rough.

The skin and the subcutaneous tissue appear distinctly thickened over the knee and leg. A distinct hypertrophy is also evident in the muscles of the calves, especially on the left side. The anterior muscles of the legs are also well developed.

The increase in the size of the feet is remarkable. Their skeleton is hypertrophied in all its dimensions. It is much the same with regard to the toes as with the fingers. Sensation and reflexes normal.[1]

The bones of the pelvis seem very thick and massive.

There is increase in the area of cardiac dulness to the breadth of three fingers outside the maxillary line. A systolic bruit is heard over the left side of the heart and at the lower end of the sternum, diminished in intensity above.

No changes are detected in the liver, spleen, or lungs. The urine contains a very small quantity of albumen, with some hyaline and granular casts.

The labia majora are somewhat large; the neck of the uterus cannot be felt on account of the thickening of the soft structures around the external genitals.

Temperature of body absolutely normal.

Nothing noteworthy with regard to the central nervous system, nor with the organs of sense. With regard to the eye, $3 = \frac{6}{12}$. The crystalline lens is a little opaque, also some floating bodies in the anterior part of the vitreous body. No notable alterations in the fundus of the eye; the retinal vessels are not dilated, and there is no disease of retina.

Examination of the faradic cutaneous sensibility by Erb's method gives normal reactions on the face, upper and lower

---

[1] Measurements of feet given.—TR.

limbs, except on some parts of the right thigh, where the strongest faradic currents fail to cause any pain, whilst only the slightest sensation was produced with constant changing of the poles. Here Professor Erb gives a table of the different electrical reactions.

The results obtained showed a very feeble electrical resistance of the skin; and, on the other hand, the intensity of the currents necessary to excite the muscles and nerves was extraordinarily great. Dr. Erb does not wish to assert that this is of a pathological nature, the age of the patient, the thickness of the skin, and also that of the subcutaneous adipose tissue, being taken into account; but thinks that, at any rate, the fact should be noted.

During his stay at the hospital the patient had two attacks of migraine, with frequent and excessive vomiting. The treatment adopted as a trial during her stay, Dr. Erb states, consisted in galvanization of the head, of the nape of the neck, and of the sympathetic; also tepid baths, lactate of iron, and extract of Peruvian bark.

When the patient went out she seemed distinctly better; she could use her hands more freely, and they seemed softer and less gigantic.

In October, 1887, Dr. Erb got this patient to return, in order to see if he could find any dulness at the upper part of the sternum, which had been noticed in the two brothers Hagner (cases published previously by Friedreich, and of whom mention will be made later in this work). The following were the results of his examination. The upper half of the sternum, and that part of the thorax immediately adjoining it, were of distinctly diminished resonance. The area where this dulness is present forms a trapezium, of which the upper border measures about 12 centimetres, the lower 8 cm., and the sides 9 cm. The cardiac dulness at the upper part extends into the lower limits of the area above described. It extends on the right side to the right border of the sternum. On auscultation a loud systolic bruit is heard, extending from the middle of the area of cardiac dulness to the clavicles. The heart sounds are somewhat irregular, and its pulsations intermittent. No arterial thickening can be detected. A careful examination of the urine gives the quantity in the twenty-four hours as about 1300—2250 cubic centimetres. Specific gravity 1017—1023, slight traces of albumen, abundant phosphates. Somewhat frequent shortness of breath. The patient has also experienced pains in the head, with vomiting.

At the end of ten to twelve days the cardiac pulsations were regular; the systolic bruit had become more feeble.

Case XXI (Erb). Shown at the Medical Congress at Heidelberg.—Mdlle. B. N—, æt. 25, seen August 7th, 1889. She had formerly consulted me February 5th, 1886, when she complained that for nine months she had experienced weakness, dyspnœa, palpitation of the heart, headache, and loss of appetite. The menstruation had completely ceased for ten months. Insomnia; melancholia; irritability. Objectively nothing could be found beyond very marked anæmia. The heart and internal organs were normal. The features were swollen. The skin of the hands was distinctly thickened. She was a person of very tall and strong build. I made the diagnosis of chlorosis, and added with a mark of interrogation, commencing myxœdema; acromegaly was not then known.

At the next visit, the patient stated that after the first examination and the treatment then prescribed (iron, quinine, and nux vomica), she was much better for a time. For the last eighteen months she has become worse.

The head and the nape of the neck are stiff; the patient cannot read on account of pain in the eyes. Menstruation is almost completely absent (it has only appeared twice during the last year, 1889). Weakness, sleeplessness, feeble voice, no palpitation, appetite and motions regular. She complains of a pain situated at the base of the cranium; each step she takes causes headache, and she cannot see well. She has a sensation as if something was behind the eyes. No trouble with the ocular muscles. Objectively: face enormous, large nose, lengthened lower jaw. The lower set of teeth project forwards. The lips are very thick. The alveolar process of the upper jaw is distinctly thickened. The tongue is enormous. Hands very swollen; fingers enlarged. She cannot easily obtain gloves which will fit.

The bones of the forearm and clavicles are very much hypertrophied; the body of the sternum and the first ribs form a very distinct projection. There is a distinct triangular area of dulness behind the upper part of the sternum. A goitre is also present, principally on the right side. Anæmia. Heart normal.

No similar affection in the family.

Case XXII (Hadden and Ballance).—A married woman æt. 35. She had had three children in ten years; no miscarriages. Nothing noteworthy as regards the family. Up to the last two years and a half she had never been ill. At this time scarlatina appeared in the house, and one of her children died of it. The patient herself had a bad throat, and œdema of the feet, but no eruption. She stated that there had been œdema of the hands at the same time, but not of the face. Following the scarlatina she had a painful swelling of both knees, which resembled an attack of rheumatism. This woman attributes her present illness to the attack of scarlet fever, although she recollects that she had prickings in the hands previously to it.

Menstruation had ceased some months before the scarlatina, and has not reappeared since.

On examining the face it is found to be swollen and puffed, differing in a very marked manner from a photograph taken some months before the scarlatina. The nose is enlarged, and the alæ thickened. The upper and lower jaws are uniformly increased in size. The lower lip is thickened, red, and everted. The submaxillary glands are easily felt, but are not hypertrophied in any definite manner. The neck is full, short, and the subcutaneous tissues seem to be over-developed. The dimensions of the thyroid body were normal; it was slightly prominent above the clavicles. The clavicles themselves were much curved, and distinctly wider and thicker than normal. The hands were remarkably large; their increase in size being due to a thickening of their subcutaneous tissue, which was particularly marked at the inner border of the hand. The skin was moist; there were folds on the backs of the fingers. The nails were large, but otherwise appeared healthy. The phalanges and metacarpals were not enlarged.

The circumference of each hand taken at the middle was nine inches; formerly the patient wore gloves No. 7.

The feet were more affected than the hands. Their circumference at the middle was 12 inches. Before the malady the patient wore shoes No. 4 size; she then took to No. 6 size, and now has to wear No. 8.

There is a very marked subcutaneous thickening at the outer border of the foot on the plantar surface, and above the inner malleolus; but very little on the back of the foot. The hypertrophied parts form masses which can easily be held between the thumb and index finger. The hair is smooth and of normal appearance; there is no tendency for it to fall out. The heart and lungs are healthy; the urine is not albuminous.

The speech was not slow, but slightly guttural; which was attributable to an unusual hypertrophy of the tonsils, which interfered with speech, deglutition, and respiration. In fact, it was for this that the patient asked the advice of Dr. Ballance, who removed the tonsils. This woman enjoyed fairly good physical and mental activity. The skin of the body was natural, as well as the sweat secretion; there was no anæsthesia. At the meeting of the Clinical Society of London on April 13, 1888, Messrs. Hadden and Ballance made further observations on this patient. These notes were taken at an examination of the woman, made eighteen months after the first (October 13, 1886):—The thyroid body is not felt on palpation, but it cannot be definitely asserted that it is atrophied, on account of the thickness of the neck causing difficulty of examination.

There is no sign of hypertrophy of the thymus, especially no dulness behind the upper part of the sternum. There is a little swelling above the clavicles; these bones are more curved and distinctly larger and thicker than normal.

Circumference of the hand opposite the metacarpo-phalangeal joint—right hand $9\frac{3}{4}$ inches, left hand $9\frac{1}{2}$ inches. The right middle finger is $4\frac{1}{2}$ inches long; the left, $4\frac{1}{4}$. The increase in circumference of the hand may be said to be $2\frac{1}{2}$ inches. No weakness of muscular force, the patient can use her hands as well as formerly. The circumference of the right foot at the base of the toes is $10\frac{1}{2}$ inches; opposite the middle of the sole of the foot, 11½ inches. The forearms and legs show no increase in size.

The speech is not slow, but somewhat guttural. Tongue large, probably hypertrophied.

The patient is almost completely blind with the right eye. The date of diminution of vision is uncertain, but the patient believes that her sight has been affected since she had what she calls diphtheria, three years ago. Four months later she discovered that she could hardly see. No recollection of headache on the right side.

Mr. Nettleship made an ophthalmoscopic examination, and recorded the following:—" There is little reaction to light in the left pupil, but accommodation is good. Bluish-white atrophy of the right disc; its veins are very tortuous. There is a small spot of pigment in the right choroid, a little above the disc. Arteries normal. No other appearances of neuritis. Left pupil normal."

Mr. Nettleship considered that there had probably been a neuritis of the right optic nerve.

No excessive perspiration or increased thirst. No curvature of the spine. The head does not approach the sternum.

The cranium has not undergone any alteration in size; the patient has never been obliged to change the size of her bonnets.

Case XXIII (Godlee).—A woman, æt. 41, consulted Mr. Godlee for an increase in the size of the thyroid gland, of nine years' duration. A cyst had formed in it which had caused neuralgia by pressure on the branches of the cervical plexus. Mr. Godlee opened and drained this cyst with amelioration of the symptoms. The patient, who had formerly been of a slight figure, and had possessed a good voice, noticed at first disappearance of high notes, then swelling of the neck, and the sudden arrest of menstruation at the age of thirty-six. Since this time a gradual increase in the thyroid body has taken place, accompanied by an increase in size of the bones of the face and limbs, and especially of the lower jaw, the hands, and the feet.

The patient was of a gouty and rheumatic family, and had been subject to rheumatism before this malady, but not since.

Present state.—Lower jaw much increased in size, so that the teeth of the latter cannot be adapted to those of the upper jaw. The lines of the head were little if at all altered, so that the face presented the shape of an egg with the large end below, thus differing considerably from the appearance it takes in osteitis deformans. The clavicles and the ends of the ribs were massive, so that the sternum appeared as if sunk in. The bones of the limbs were not generally thickened, but all their natural prominences were much exaggerated, and the small bones of the hands and feet were very much increased in size; these extremities themselves were also large, and in the shape of spades. Marked kyphosis, suggesting spinal caries, and producing a considerable diminution in height. The cartilages of the ears, and probably of the nose and larynx, are thick and resistent. Skin thick, with large sebaceous glands on the face; otherwise natural. Subcutaneous cellular tissue normal, but deficient in amount on account of the emaciation. Profuse perspiration; formerly the skin did not present any abnormal moistness. Muscles much atrophied. Hearing normal; smell decidedly affected, especially for delicate odours. The same also with taste; the tongue was very thick and large. Sight good. Voice harsh, metallic, and monotonous. A little dyspnœa, due evidently in part to increased size of the thyroid gland. The general appearance is that of marked weakness; the patient walks dragging her limbs like an old woman. Appetite regular, but excessive thirst. Pulse rapid. The urine contains no albumen nor sugar. Intelligence perfect, and temper placid.

Case XXIV (Wilks).—The subject was a woman of 28 seen by Dr. Wilks (1869), and of whom he has only recorded a few notes. She was of good features before her illness, but since has become so hideous that the boys cry after her in the street. Her features were thickened and deformed; the hands were large and clumsy. The malady had progressed for six years, during which amenorrhœa had persisted. She had lost the sight of both eyes. She had been given arseniate of soda, which seemed to do her good. In April, 1870, she went into the country, and was reported to have died of coma. The opinion of Dr. Wilks at this time was that the woman had a "cerebral tumour, this diagnosis being based on the persistent headache and optic neuritis." It was on account of her persistent headache that the patient had consulted Dr. Wilks. He was of opinion that the malady "was entirely distinct from all others," and that these cases belonged to a class totally distinct from those of ordinary pathology.

CASE XXV (Tresilian, 1).—E. M—, æt. 31, had rheumatic fever between twenty and twenty-one years of age, and had been told that she had "cardiac dropsy." Her present malady dated for about six years. The swelling was the first recognised symptom; it commenced in the hands, and showed itself some time after in the face around the eyes. Her menstruation was regular up to the time when the swelling commenced, but has not reappeared since. The face is large and broad, waxy and pale, as in a patient the subject of chronic Bright's disease. There is very marked swelling of the lids, especially of the upper. The mouth is large, the lips swollen. The tongue is large and flabby. The speech is a little slow, but distinct. The hands are large and much swollen. The fingers are much thickened over their dorsum (but do not pit on pressure). The feet are also very swollen; the patient cannot get into the shoes she formerly did. The nose is very thick, particularly at the alæ, which are of a purple colour. The ears are very large and swollen. The swelling of all these parts has supervened gradually, and position appears to exert no influence, for she remains about the same morning and evening. A rotatory "nystagmus" has also been observed at times. There is nothing abnormal in the fundus of the eye.

CASE XXVI (Minkowski).—A Russian musician, æt. 30; no hereditary antecedents. Married at the age of twenty, and has had eight children, who are well developed, although somewhat feeble.

In 1887, at the age of twenty-eight, it appeared to the patient that his fingers began to grow thicker. He was obliged to leave off a ring, which he had habitually worn, because it had become too small. He did not take any further notice of it. Two years later he suffered from continuous pain in the head ; soon after he noticed that his feet began to grow larger. His boots became too narrow ; and instead of No. 9, he had to take to india-rubber shoes of Nos. 10, 11, and finally 12 size. The hands increased also in thickness more and more, so that the patient could no longer play the violin, it being impossible to give distinct notes. He began to play a cornet, but was soon obliged to use a larger mouth-piece, because the lips had themselves also become thicker. He had noticed besides more recently a distinct thickening in the nose and ears. The general shape of the face had changed during later years. There was increased pain in the head, which, although continuous, presented paroxysms on the left side. During the summer of 1886 there was diminution of sight ; at first in the left eye, then in the right. This diminution became such that, when playing, the patient could not read the notes. Hearing also seemed diminished in the left ear.

Present state.—November 1, 1886. The patient is of medium height (164 centimetres) ; looks anæmic, but is fairly well nourished. At first glance one is struck by the extraordinary size of the hands. These appear enlarged and thickened throughout, a little short in comparison with their width. The fingers are equally enormously thickened. The increased size is manifestly due in part to a thickening of the bones ; but the hypertrophy of the soft structures also plays a still more important part, so that the normal projections at the articular ends of the bones are but slightly marked. The subcutaneous cellular tissue as well as the skin is hypertrophied throughout, but seems otherwise to present a normal appearance. The epidermis is absolutely normal and soft. The nails are enlarged, a little flattened, and present a distinct longitudinal striation ; in comparison with rest of the fingers they appear rather small.

The forearms appear also a little thickened, though to a less marked extent than the hands. There appears, at any rate, a distinct disproportion between them and the arms, of which the dimensions present little abnormal. The whole upper extremity shows the type of an hypertrophy increasing from the base towards the periphery. As regards the lower extremities it is the same as in the upper; there exists in them also a definite hypertrophy of the peripheral parts. The thighs appear normal; the legs present a moderate increase in size; as regards the foot, it is enormous, and also the still more distal parts (the toes and heel), which show a great increase in size. The great toes have an appearance really almost gigantic. The patella seems a little thick and massive, as well as the rest of the knee, which is somewhat large.

On the face there is also distinct hypertrophy, especially marked on the nose, lips, and chin. On the nose in particular the tip and the septum are much increased in size, as shown in Fig. 93. The lower jaw, especially at the chin, is manifestly thickened, a little widened, and considerably lengthened. The teeth of the lower jaw are thus placed a little in front of those of the upper. The soft structures are equally thickened over the chin, which is strongly projecting. The zygomatic bones and orbital ridges of the frontal are also very distinctly prominent. The entire head has the peculiar elongated oval form, which corresponds exactly to that which P. Marie has described as characteristic of acromegaly. The ears themselves are remarkably large and deformed, and the cartilages are distinctly felt to be of an abnormal thickness and resistance, and the auricle cannot be folded completely forwards.

The eyes are somewhat singularly prominent; in the left especially there seems to exist a true exophthalmos; the palpebral fissure is not, however, widened. The tarsal cartilages seem a little enlarged. The pupils on both sides are equally small, reacting equally to light and accommodation, although in a somewhat sluggish manner. The ocular movements are well performed.

The tongue is broad and thick; the mucous membrane of the cheeks and palate is specially hypertrophied.

The cartilages of the larynx appear perhaps a little enlarged, especially on laryngoscopy, but without it being possible to establish a definite hypertrophy.

The thyroid body seems distinctly atrophied; it can only be examined with difficulty.

The vertebral column presents a uniform kyphosis in the upper part of the dorsal region; the head is strongly inclined forwards. The soft structures of the neck are a little thickened. The ribs do not seem to be enlarged; the intercostal spaces can be distinctly felt. The sternum itself is not hypertrophied. The xiphoid appendix is distinctly enlarged and thickened.

The muscles of the entire body are badly developed, and somewhat soft. There exists no paralytic manifestation, but the patient's strength is generally somewhat feeble. Power of the left hand with the dynamometer is 20 K, right 23 R. Walking is a little laboured, especially so on account of the foot having a tendency to fall a little downwards. Nothing special to note in the joints.

Sensation is everywhere normal; also the skin and tendon reflexes. No trouble with the urine on micturition. Minute examination of the vision gives considerable diminution of sight, especially in the left eye (fingers cannot be recognised further than 5 cm.). The visual field is diminished in the right eye on each side; it is totally lost in the upper and outer segment. In the right eye there is equally total loss of vision in the upper and outer segment to within 15° of the fixation point. The visual field for colours has not been tested in any special manner. The patient recognises green with both eyes. Ophthalmoscopic examination gives normal results, without any difference between the right and left eye. The acuteness of hearing is also more affected in the left than the right ear. The patient hears the tick of a watch with the right at a distance of 20 centimetres, with the left only at a distance of 5 centimetres. Taste and smell are normal.

Examination of the internal organs reveals nothing abnormal; the urine presents the usual constituents and quantity. No increased thirst. The patient's intelligence is good; speech is slow and a little monotonous, but presents no other anomaly.

The patient is somewhat depressed in spirits; which is to be explained by the persistent headache, by the progressive diminution of sight and hearing, as well as by the difficulty with which he follows his occupation.

CASE XXVII (Wadsworth).—Mme. C—, æt. 42, a strong and well-built woman, consulted me at the Boston City Hospital (on June 27, 1884), on account of diminution of sight. She was in fairly good health, and since measles, which she had at fourteen, she had never been ill. She had been married twelve or fifteen years, but had had no children. Her menstruation had been regular till three years ago, at which time it ceased. About seven years ago her fingers and hands, and two or three years later her feet, together with the lower part of her face, nose, and eyelids, commenced to increase in size. This increase developed progressively, and for a long time she was obliged to have her shoes expressly made, being unable to find any sufficiently large. Her speech became a little thick and slow; her movements also slower than formerly. Except for transitory sensations, numbness in the parts affected, and burning in the feet and legs, she has never experienced any real pain. Recently, however, she has felt a slight dyspnœa after active exercise. The diminution of the sight was not noticed till about a year and a half ago, and has constantly progressed. A careful inquiry failed to elicit the existence of any pain. She has never had any headache. There was no diminution of intelligence or memory. The appetite was good, the bowels a little sluggish; micturition normal; sleep heavy. The lower part of the face was full, heavy-looking, and of a waxy tint; the natural folds had disappeared.

The lips were thick; the nose large, its alæ broad and thickened. The lids swollen like little bags. The fingers and hands were swollen and square in shape. The thickening was specially marked on the palmar surface, where the tissues could be pinched up into thick folds, giving the sensation of excessive development of the subcutaneous adipose tissue. The feet were very thick and broad. The swelling did not extend above the malleoli, and although there were large varicose veins on the right leg there was no œdematous pitting, either there or elsewhere. The tongue was large and somewhat pale. The skin was not dry to the touch, but the patient stated that she never perspired. The thyroid was of normal size. Auscultation revealed nothing abnormal with the heart or lungs. The temperature and pulse were normal. The patella reflex was present. No anæsthesia of skin. Urine normal in every respect. On examination of the eyes, the conjunctiva, cornea, and iris were normal. The ocular movements were well performed. With the right eye light alone was perceptible, with the left vision was 20 (P. contracted in all directions, but to a much greater degree upwards and outwards). The media were translucent. In both eyes the discs were distinctly defined, grey with a slightly bluish tint, without being vascular. The central vessels, arteries, and veins were small; in other respects the fundus was normal.

CASE XXVIII (Fritsche and Klebs).—Peter Rhyner, æt. 44, a bachelor. No analogous affection has been observed in his family. He presented at birth all the appearances of normal health; no rickets or other malady. During the first years of adult age he was considered of medium height among the people of Elm (the inhabitants of this country are distinguished for their height and fine stature). The patient did not know exactly what he measured, but he recollected that when he was a soldier he was little less than six feet. His back was at that time perfectly straight. His weight was 75 kilogrammes. The present malady commenced about eight years ago, that is at the age of thirty-six years, by pains and weakness in the hands. The pains did not restrict themselves to the joints, but attacked all parts equally. They are described by the patient as stretching and tearing. At the same time there was a slight increase in the size of the hands, with redness; from the description of the patient there was no acute inflammation nor true œdema, for it did not pit on pressure.

Little by little the pains went to the arms, and showed themselves, though less severely, over the whole body. The pains in the head were pretty severe and frequent. For the last six years they have been located behind the ears. During the last three years they have extended, with less intensity, over the whole head. The face has always been a little pale. It is only during the last two years that the legs have become painful, particularly the knees this summer, but the feet have never been distinctly painful. During some time the patient fancied that he had prickings in the hands and diminished sensation, but these symptoms have all disappeared, as well as the very marked tendency to perspire in the hands. At the same time as these pains, the size of hands, feet, ears, lips, and also of the whole head, neck, and knees, began to increase. The hypertrophy of the fingers is the most striking; and also of the last-named position, that of the knees. The increase in size appeared to supervene everywhere at the same time. The circumference of the thorax has also increased in a very marked manner, but according to the patient the arms and legs have not undergone any increase in length. It was also at this time that a curve in the vertebral column showed itself—a slowly progressive kyphosis.

For about two years the increase in size of the different parts of the body and the kyphosis have been arrested; in fact, during the whole year M. Fritsche has observed him they have undergone no change.

About the same time as the other symptoms weakness supervened, which increased in a few years to such an extent as to make walking for even a few minutes very laborious. For four or five years he has had dyspnœa on the slightest effort, and for six or seven years frequent palpitations.

The patient has never been a great eater. Constipation has been present for a number of years. The quantity of urine has not been noticed to be abnormal. For many years there has been total loss of all sexual function. The patient has noticed a diminution of the memory. The eyes have become weaker, but hearing and the other senses are well preserved.

At first glance, the increase in the head, the hands, and the feet make a striking contrast to the short length of the body. The latter is explained by the existence of the marked kyphosis, which has brought about shortening of the body. The face, seen from the front, is too large, compared to the rest of the cranium, in consequence of the great kyphosis between the shoulders. The face and specially its lower part, particularly the nose, lips, and chin, are remarkably prominent. The dimensions of the outer auricle seem also excessive.

The penis is large, but not to any extraordinary extent.

The scrotum and testicles are certainly atrophied. The legs and arms are of a slimness which contrasts strongly with the size of the feet and hands. The knees are distinctly thick in comparison with the thighs and calves.[1]

It is seen from the different measurements that the bones of the face are increased in size; the forehead is very low in comparison with the other parts of the face. The width of the malar region is increased, without the malar bones appearing markedly prominent. The lower jaw is increased more in its antero-posterior direction than in the transverse. Though the palate seems high and narrow, the distance between the upper canines is, however, greater than in an ordinary individual. The ears, nose, lips, and tongue show a very marked increase in size.

As regards the trunk, besides shortening due to the kyphosis, it is necessary to note the very marked depth of the thorax (antero-posterior diameter), whilst the transverse diameter is rather small. The sternum is of very considerable length; the scapulas and clavicles (?), on the contrary, are not increased. The size of the pelvis is certainly larger, even allowing for the thickening of the integuments.

As regards the upper limbs the arms do not present any increase in length. The bones of the forearm are a little longer than those of an ordinary person, but it is only the hand which presents any important difference. In comparing the different segments of the upper limbs, it is found that " the arms of Rhyner are of less size, that his forearms have the same size, and that his hands and fingers are out of all proportion " to those of a normal person.

The patient has all the appearance of very marked anæmia. The integuments are pale, a little dry, and somewhat dark. No pigmented spots; no varices or œdema. No elephantoid infiltration of the skin, nor any warts. On the left calf is some chronic eczema, but nothing of importance. On the back there are some outgrowths of molluscum fibrosum. On the abdomen, which is short and a little retracted, the skin is fat, and is divided by deep transverse folds, one of which crosses the umbilicus. On the palms of

[1] Detailed measurements of the different parts of the body given by MM. Fritsche and Klebs.—TR.

the hands and soles of the feet the skin is thick and resistent, and can be raised into large soft folds. It appears to be hypertrophied throughout its thickness, and the subcutaneous tissue seems to be equally if not more so. This hypertrophy is less marked at the elbows and knees. The skin appears as if too large; it is freely moveable, and can be raised in great soft folds much more easily than usual. The abnormal thickening of the skin is specially noticeable over the occipital region. Over the outer occipital protuberance is a horizontal exostosis in the form of a comb, about 3 centimetres long and 1 centimetre high at its centre (probably congenital). Above it near to the top of the head are four or five longitudinal ridges about the thickness of a finger, and the length of 12 centimetres. Above these ridges flattened bony prominences are found, reaching from the parietal bones to the occipital.

The hair is abundant and thick; it grows with great rapidity. The hairs on the body are well developed. The beard and hairs of the pubes are abundant.

The nails are distinctly enlarged, but in comparison with the fingers of the patient they appear rather small. The small phalanges of the fingers are comparatively less developed than the other segments of the fingers. The nails grow rapidly.

The muscles are particularly soft and small, notably those of the extremities; only those of the neck are thick and firm. The muscular force is much diminished. The atrophy of the muscles is specially marked on the arms, the calves, and also particularly on the thenar eminences and back of the hand, where, on account of disappearance of the interossei, deep furrows are seen between the metacarpals.

Among the arteries accessible to the touch, the femorals and brachials present a certain degree of rigidity, but the radials and temporals are soft. The pulse is small and feeble. The number of pulsations is somewhat freqnent (48 to the minute). The inguinal glands are a little large, but there is nowhere else any gland enlargement.

The teeth are separated one from another, less in front than behind, and this appears to be much more so than formerly. They are of normal size. When the jaws are closed the lower teeth project in advance of those of the upper jaw. The larynx and hyoid bone do not seem increased in size.

As regards the knees, the articular ends are thickened, independently of hypertrophy of the skin. There is no exudation in the joints, but in both knees there is crepitation when movements are performed, like in arthritis deformans.

In the scapulo-humeral joint there is also dry grating. Opposite the spinal curvature the three or four first dorsal vertebræ (spinous processes) are nearly on the same level, that is to say, directed horizontally backwards. Below, the spinous processes are arranged in such a manner that the point of greatest prominence posteriorly is opposite the ninth or tenth, from which it descends nearly vertically to the sacrum with only a slight degree of lordosis. The mobility of the spine, except that of the neck, which is quite free, is only demonstrable in the lumbar region, and is there very limited.

The curvature is almost entirely a simple kyphosis; there exists only a slight degree of deviation towards the right (vide Fig. 95). Corresponding to the kyphosis, the sternum is in its lower part distinctly projected forwards, so that its direction is very obliquely forwards. It is only at its lowest part that it is vertical for a very short extent, being there continuous with the xiphoid appendix, which is strongly bent inwards.

The ribs are, as in curvatures in general, directed perpendicularly downwards. The lower border of the ribs almost touch the iliac bones. Their extremities where the cartilages join are at one part very flat, and form a very distinct chaplet of beads.

Nothing special in the mouth or larynx. On both sides there is slight increase in the lobes of the thyroid body. Cardiac dulness, upper border at the second and fourth ribs, left border from the sternum to the width of a finger outside the nipple. There is considerable impulse at the apex in the region of the nipple, and just two fingers' breadth outside it. Systolic bruit distinct at the apex, feeble at the base. Diastole clear. Heart's action feeble, but regular. The dulness over the spleen is not found to be increased. Hepatic dulness, upper border of the fourth or sixth rib. The patient coughs and expectorates a little ; moderate secretion. The urine presents nothing particular, no albumen, no sugar. Constipation.

Intelligence is intact. The patient has also an excellent temper. No diminution of memory is perceived in his conversation. The eyes are hypermetropic, but of normal acuteness of vision. The ocular globes are increased in size. Slight arcus senilis. The cutaneous sensibility is normal to a prick and to touch, it seems diminished on electrical examination. Hearing, smell, taste, perfectly intact, but electrical excitability is diminished for the faradic and galvanic currents. In spite of the hypertrophy of the tongue, pronunciation is good, speech is, perhaps, a little slow.

On September 17th the patient experienced malaise, fainting, loss of appetite, with weak pulse. On the 18th the same symptoms, pulse small, very slow ; three more fainting fits, of which the last terminated in death.

Necropsy (a short *résumé* of this patient).—Cutaneous covering of the cranium uniformly thickened, 9 millimetres to 1 centimetre. Subcutaneous cellular tissue loose, without increase of fat. Periosteum fairly normal and easily separated from the bones. The cranial vault is of an elongated oval shape, with narrowed frontal portion. It is distinctly thin at the sides (temporal fossæ). Occipital portion distinctly thickened. The same also with the parietal part near to the sagittal and coronal sutures. The coronal sinuses are remarkably large. The sagittal and frontal sutures are indistinct, and at certain parts ankylosed. The inner surface of the bones of the cranium are smooth throughout, but grooved by numerous deep vascular channels. On the inner surface of the left parietal bone are numerous flattened osteophytes (osteo-periostitis ?) which adhere to the dura mater.

Longitudinal sinus large. Pia mater smooth and thin. The cerebral convolutions are very much convoluted, with numerous folds; the fissure of Rolando only shows clearly at the origin of the fissure of Sylvius. The entire brain is evidently uniformly increased in size. The cerebral cortex is not thickened, the lateral ventricles are a little enlarged and filled with a clear serum. The arteries at the base of the brain are distinctly enlarged; as regards the vertebrals, the right only is much developed. The basilar is particularly large, and its walls uniformly thickened.

The cranial nerves in general are more developed than normal. The increase in size is specially noticeable in the third pair and optic nerves. The latter are compressed laterally by a tumour the size of a hazel-nut, which fills the sella tursica, forming a hemispherical swelling inside it.

The right optic nerve is a little more compressed than the left, being distinctly flattened and separated from the tumour. It presents the form of a flattened band a centimetre in width. The left optic nerve is more rounded, its width is 6 millimetres. The trunk of the infundibulum is distinctly increased in size, and placed in the form of a large yellow band against the right side of the tumour.

The pituitary tumour, extirpated with part of the sphenoid, presents an extremely soft mass, liquid in the interior. It is enveloped in a thin membrane, rich in vessels, into which the infundibulum is prolonged. The posterior wall of the sella tursica is altered into a thin bony plate, with a normal investment of dura mater. In front, the plate of the ethmoid, which is unchanged, limits the tumour. The floor of the sella tursica is unusually deep; there is no communication between it and the cavity of the sphenoid.

The pharyngeal mucous membrane is much swollen, congested, and covered with viscid mucus, resulting in considerable narrowing of the aperture of the pharynx. Its pillars are much hypertrophied, covered with a dull red mucous membrane, and forming a distinct projection. The mucous covering of the nasal fossæ does not seem changed.

The globe of the eye seems increased in size.

The costal cartilages are of extraordinary width and thickness, but of regular shape. At the point of their insertion into the bones they are enlarged, which gives a little the appearance of the chaplet of beads in rickets. The pneumogastrics in the neck are uniformly thickened, almost to the size of the normal median. The sympathetics are also a little larger than normal. The carotids are extraordinarily large, with uniformly thickened walls.

The heart is large, and very soft; the mitral and tricuspid admit three fingers; its weight is 550 gr. Its cavities are very large; the muscular substance strong; and the papillary muscles much lengthened.

The arteries have a very much increased diameter (the authors insist much on this point and enter into considerable detail).

The remarkably developed thymus forms a triangular mass, prolonged upwards in the shape of a band as far as the thyroid.

The larynx, thyroid body, and tongue are all uniformly increased in size; but without other alterations. The glands at the base of the tongue are a little swollen. The spleen is distinctly increased in size. Its weight is 750 grains, consistence firm. The left kidney is very large, the right is a little smaller, weight 275 grains. Suprarenal capules very thin, their cervical substance consists of only a thin yellow covering. The medullary substance is a little more abundant. Liver large and heavy, 2800 grains. Portal vessels large and filled with dark blood. Stomach very large, mucous membrane much corrugated. Mesenteric glands somewhat increased in size, and very red. In the small intestine much mucus; its follicles swollen. Bladder and prostate not changed. Testicles of medium size, pale, and soft.

CASE XXIX (Ellinwood).—"This malady is rare, and does not as yet figure in our manuals; we owe its name and its description to Sir James Paget.[1]

"The case which I present for your examination relates to a man æt. 28, who had been delicate during adolescence. No known hereditary antecedents. At the age of twenty-one (1876) he weighed 93 pounds; now (April, 1883) he only weighs 196 pounds. He has always been able to work, although he experienced from time to time a little muscular weakness, malaise and lancing pains in the lower limbs, especially the knees. Mastication of food has been very troublesome; also dyspepsia. For about a year he has perceived that his lower jaw was advancing in front of his upper, and that it increased

[1] We owe our best thanks to Dr. Mibierge for our knowledge of this case.

gradually till he was no longer able to bite his food. His lower teeth were extracted and replaced by a dentist. This summer about the same time they were substituted by a second set, and as the lower jaw continued to lengthen his upper teeth were extracted, and replaced by an artificial plate. This was advanced in front of the border of the upper jaw, so as to correspond with the lower one. The deformity, however, progressed and the lower teeth became separated from the upper to the distance of about 22 millimetres. The bones of the face are also seen to be increased. The cranium is enlarged so much that the patient is obliged to wear a larger hat. The bones of the trunk, sternum, and ribs, are also enlarged and thickened; and the vertebral column presents a lordosis. For the latter he wears a corset, from which he derives some benefit.

"The bones of the hands, the digital phalanges, are seen to be much thickened, strong, and resistant. The skeleton is no doubt affected throughout; however, it is in the head and short bones that the hypertrophy is most manifest.

"When he gets up from bed in the morning his joints adapt themselves with some difficulty, producing gratings, and the patient has to manipulate them before they are comfortable. He is anæmic, and suffers much when it is cold.

"The white appearance of his hands and the pallor of his lips have diminished since his nutrition has improved. He presents no cerebral symptom nor headache; his mind is clear and his memory good. This man has never had any articular rheumatism. No sugar nor albumen; quantity of urates increased. As far as can be judged the bones are not porous, and their weight is not diminished. Their enlargement is symmetrical. The vertebral column is bent, its natural curves being much increased. Sir James Paget points out the characteristic symptoms of some of the other diseases of bones which distinguish them from osteitis deformans. In rickets, for example, the bones are very small and very short, and the curves which the thickened and lengthened bones present are in various directions. In osteomalacia they are thin and curved in an angular manner.

"The treatment which I have adopted for this patient has been dietetic, believing that we have in the bones a large quantity of histogenetic substances.[1] I allowed him a very moderate quantity of food, and prescribed cod-liver oil for the respiratory organs.

"The digestion is better. The anæmia is less marked, and the general health is better. I have advised him to stay in a mountainous district, and to see me again on his return."

CASE XXX (Taruffi).—Marchetti, a cooper by trade, died at the age of 47 in 1808, after having presented unmistakable symptoms of acromegaly. In 1879 M. C. Taruffi, Professor of Pathological Anatomy at Boulogne, described in a very detailed manner the skeleton of Marchetti, which was preserved nearly intact in the pathological museum of that town. Taruffi, however, was of opinion that the skeleton had belonged to a giant. This theory has, however, been demonstrated to be a mistake by M. P. Marie, who has subjected the work of the Boulogne professor to a critical examination, showing beyond doubt that Marchetti, far from being a giant, was clearly the subject of acromegaly.

M— was known as a great eater; his height was 1·80 m., His face was very long in the vertical diameter; a change in which Taruffi states the cranium shared to a very slight extent. His lower jaw was increased particularly from behind forwards. Its angles did not exist, but were replaced by two curved lines; in short, prognathism of the lower jaw. The extremities (hands and feet) of Marchetti were those of acromegaly (see the report of M. P. Marie).

In the absence of details on the condition of the internal surface of the cranium in the description of M. Taruffi, Marie states that he wrote to his distinguished colleague and friend Dr. G. Melotti (of Boulogne), and requested him to fill up this deficiency by examining this part of the skeleton. Below are the results of his examination, which he desired me to publish in July, 1889: "The sella tursica is considerably increased in length, width, and depth; and this at the expense of the sphenoidal sinus, which has almost entirely

---

[1] The histogenetic substances are those which serve for the molecular repair of the normal tissues.

disappeared, as well as the depression which receives the commissure of the optic nerves."[1]

CASE XXXI (Brigidi).—Ghirlenzoni, who was well formed in his youth and of good proportions, (having a liking for it) had entered the theatrical profession. The following notes are from an account of him by Michel Casini, barber at Florence :—" I have known Ghirlenzoni since 1835. He was then a young man of about twenty-five years of age; his features were perfectly regular, and the different parts of his body showed no disproportion, except the nose, which in length distinctly exceeded the ordinary dimensions. He married a young widow, and had one daughter still living and well formed. At thirty-five years Ghirlenzoni was in Italy, and there contracted syphilis. When he returned to Florence in 1859 he was already deformed. He stated that he had suffered for more than two years from this terrible disease, and that during it his hair had fallen out, and his body been covered with spots and pustules. From this time Ghirlenzoni had never been well, and his deformity became more marked year by year.

"He had never, according to his friends, suffered when young from rickets. It is not known at what age his deformity began; but what seems certain is that it has shown itself in a more pronounced manner as age advanced. It commenced at a time when the deformity was a means of success to Ghirlenzoni, who was still at the theatre; and Brigidi records with what enthusiasm the Florentines applauded him when he played amorous *rôles*. As for example in Francesca da Rimini, he would cry out, speaking strongly through his nose,

>   . . . t' amo Francesca, t' amo
>   E disperato è l' amor mio.

"The disease, however, progressed without ceasing, and his hump became very marked. Still, more unfortunately, he could not speak clearly, on account of the excessive size of his tongue. This organ was no longer confined inside his

[1] Measurements of the sella tursica given, showing increase in all its dimensions.—TR.

teeth, but projected beyond the incisors. It was with difficulty his legs supported him, and he walked leaning on a stick. He was, moreover, addicted to drink, and had become very short of money in order to satisfy this craving.

"Ghirlenzoni was a fairly tall man, his thorax was flattened laterally and prominent both in front and behind. The posterior curve was, however, more marked than the anterior; and was not exactly in the middle line, but inclined to the right side.

"The abdomen followed the anterior curve of the thorax, describing with it the same arc of a circle. The large and massive trunk was supported by two thin and straight lower limbs. The upper limbs were lank and long; the hands alone were broad and thick, with long and knotty fingers. The head seemed continuous with the trunk on account of the short neck, which was inclined towards the left side. It was large and flattened at the sides; his hair was scanty, short, and becoming grey. The broad forehead, which was divided by deep transverse furrows, was limited below by the two large superciliary ridges, covered with long dark hairs. This region, generally regarded as a sign of intelligence, was in Ghirlenzoni well formed. Below it, deep in the orbits, were the small dark eyes, showing but little expression, and out of proportion with the large forehead.

"The rest of the face had more the appearance of an ape than a man. It was lengthened, with very marked prognatism, flattened and indented laterally, as if the cheeks had been elevated by a blow from a hachet on each side. The thick, broad, fleshy tongue had become interposed between the widely separated alveolar borders. The lower lip was hypertrophied and everted, causing the face to be still more deformed; the nose was large and aquiline. Surrounding his face he had a thin short beard, prematurely grey.

"Being so miserable and weak, Ghirlenzoni resolved to commit suicide, and threw himself into the Arno on the 29th August, 1875. When in the water, however, he made cries for help, and some watermen were able to save him. According to their account 'he floated like a cork.' After being carried to the hospital he was seized with delirium, and died from coma on the following morning."

The autopsy of Ghirlenzoni is very detailed; a *résumé* would omit many points of importance, and we prefer to refer the reader to the account of it in the work of M. P. Marie ('Nouv. Iconogr. de la Salpêtrière,' 1889, Nos. 4 and 5).

CASE XXXII (Henrot).[1]

CASE XXXIII (Lanceraux).—In speaking of the osseous hypertrophy which is seen after the period of full growth of the skeleton, and which he thinks is due to nervous influence acting probably on the smaller vessels, our master, M. Lanceraux, gives the *résumé* of the case of a patient who he believed was the subject of exophthalmic cachexia. " The forehead," states this distinguished physician of the Pitié (p. 29),[2] " is flattened, and the bones of the cranium are manifestly hypertrophied. The nose, the lips, and especially the upper jaw, are the seat of definite hypertrophy. The teeth, which are regular at their free borders, are separated from one another, as if they had not shared in the increased development of the jaws, particularly that of the lower jaw. The nipples are well developed. The costal cartilages are for the most part ossified; the chest is convex in front, as occurs in the deformities of rickets. The thyroid body is hypertrophied. The larynx is large, and the cartilages composing it are calcified. The voice is harsh. Both upper and lower limbs are normal; the upper are relatively larger than the lower.

" *Autopsy*.—Hypertrophy of the cranial bones; brain small. Enlargement of the pituitary body, which presents the size of a small hen's egg; increase in the pituitary fossa. There is probably a communication between the sella tursica and the sphenoidal sinus. From the latter exudes a white serous fluid, with altered cell elements, and a vascular cellular membrane lines it. The brain is depressed opposite the pituitary body; the anterior lobes are small..

" The connective tissue and fat at the base of the orbit are hypertrophied. Considerable exophthalmos; eyes healthy.

[1] Given in Marie's original paper in the 'Revue de Médecine,' to the translation of which the reader is referred for the notes.
[2] Lanceraux, 'Traité d'Anatomie pathologique,' pl. iii.

Hypertrophy and dilatation of the heart, especially the left lobe; muscular fibre yellow, and more or less granular. The right lobe of the liver is almost trebled in size. There is a puckered depression, large enough to contain a pigeon's egg, on the upper part of the liver. The left lobe is very small, there are white spots on it something like the marks of former tumours; its cells are granular and fatty. Spleen large and hard. The kidneys are almost double their normal size; irregular on the surface, with little cysts. There is thickening of the walls of the stomach. Uterus very small. Ovaries the size of a pigeon's egg."

CASE XXXIV (Lombroso).—A man of 37, who had enjoyed good health up to the age of twenty-four; he then had bronchitis. When he was well again he observed his body begin to grow all at once, so that in three months he had to alter his clothes. He suffered at this time from some slight intermittent fever; and acquired also an extraordinary voracity, with some pains in the bones, joints, and stomach. His strength failed and he experienced dyspnœa and cardiac pain. Lombroso saw this man seven years after the onset of the affection. He weighed 120 kilogs. 400. His height was 1 m. 80, the skin was of a deep yellow, the beard somewhat thin. The hair, which was fairly abundant, was of a chestnut colour and coarse. The dimensions of the head are about normal; the ears are normal in size and position The face, however, is disproportioned, especially in width, and reminds one in its monstrosity of the gorilla or lion. The distance between the two cheeks is great, but still more so is the length and width of the lower jaw, which, however, in spite of its enormous development, remains nearly level with the upper jaw. The soft parts of the face do not share in equal proportions to the development of the bones. The eyes are slightly larger than normal (greatest transverse diameter 65 millimetres). The teeth are nearly all wanting; those remaining have a normal appearance. The neck is enormous, double the ordinary dimensions. The development of the shoulders, scapula, and clavicle is also enormous. The whole circumference of the thorax measures from one nipple to the other 1330 millimetres. The arm and thigh do not present any hypertrophy; but from the middle of both the leg and forearm towards their extremities the limbs become extraordinarily developed, much more so in the upper than the lower limbs.[1]

In short, the malar bones, the vertebræ, the ribs, the sternum, the bones of the forearm, of the feet and hands, are increased in size; whilst the femur, the humerus, all the bones of the cranial vault, and in part those of the pelvis have remained normal. The skin is thickened in the hypertrophied regions of the forearm, of the foot, and the face. The muscles of the hypertrophied limbs appear to the touch of a greater density than normal muscle; either lardaceous or cartilaginous.

Intelligence was good, and the patient also showed a certain refinement of sentiment. Though when young this patient had been of very amorous temperament, he has since lost all sexual desire.

CASE XXXV (Verga).—" At the end of 1860, when visiting the incurable patients at the Église Santa Maria ai Nuovi Sépolori, one of the chapels of ease of the Ospitale Maggiore de Milan, I was struck with the appearance of a patient whose face, of a pale waxy colour and disproportioned size, almost excited dread. Such, also, was the impression caused to others, for this woman was known as 'the grotesque.' Noticing that I observed her closely, she told me that she had not always been like that, but had at one time been like other girls. I asked to know her antecedents, and I received from Dr. Chiapponi the somewhat curious notes which follow: Maria B— is from Milan; her mother died at twenty-three from a malady of short duration; her father, over fifty-seven, from scirrhus of the stomach. Her brothers had died young; one only is still living, in good health, tall and thin.

"After having menstruated first at the age of eleven, these periods ceased about a year after having had variola. The menses reappeared at sixteen, and continued very abundant, but stopped again finally at the age of twenty-nine. Married, but no children. She lived by dressmaking. Since she

[1] Measurements of body given.

was a little girl she has been subject to obstinate ophthalmia, which was relieved by local treatment, but has recurred frequently during all her life. Her general health changed after the cessation of menstruation. Up to the end of this time (aged twenty-nine), during the last ten years, a great number of tumours (about fifty) have appeared in relation with the different joints, especially on the legs. They were the size of hen's eggs, of natural colour, painful to the least touch, and hard. They showed themselves two, three, or four at a time, and in two or three weeks were reabsorbed, but never suppurated. At thirty-five she got rid of these tumours, but she observed that her health was getting worse. She suffered from ascites for some time, and experienced severe pains in the legs, which rendered movement impossible. What also became still more noticeable at this time was that her limbs, which formerly were somewhat thin, began to gradually increase, so that three times she was obliged to cut the ring on her finger; her face in particular became monstrous.

"Having been admitted to the hospital for gastric fever on August 24th, 1856, she was placed among the incurables for rheumatism and amblyopia on September 16th of the same year. Since this time her pains, and particularly those in the neighbourhood of the joints, have not left her.

"When I saw her in 1860 she remained all day in bed, in which she could only turn herself with great effort and with assistance. Her intellect was a little slow, but fairly clear. Her temper had always been good, but very susceptible to the slightest contradiction. Sight feeble, hearing dull, appetite good, digestion normal. Habitual constipation; sometimes absolute for twenty-two days. It was constantly necessary to resort to measures to excite the intestines to action. Urine always abundant, sometimes passed with difficulty. The respiratory organs have caused no trouble. . . . She has palpitation from time to time, and also intermittent action of the heart, never any febrile manifestation. Her total height is 171 mm.; from the eyebrow to the chin measures about 18 centimetres; and from one angle of the jaw to the other, passing under the chin, about 29 centimetres. It was not possible to determine if the tendency to increase in size was greater in the flat bones, in the cylindrical, or in those of the mixed type.

"I find in the notes on this patient that in November of the same year she was attacked by erysipelas of the face. During the same time also in 1862 she suffered from a purulent discharge from the right ear, which also extended to the left ear; and in October of the same year she became completely blind and almost deaf. Finally, after alternations of coma and semi-delirium, she succumbed, at the age of fifty-nine, to a typhoid affection, in which large bedsores formed, and severe epileptiform seizures became manifest."—Autopsy in Marie's work, loc. cit. ('Nouvelle Iconog. de la Salpêtrière,' 1889).

CASE XXXVI (Chalk).[1]—At the age of seven, this woman, who was then in good health, fell in her shop and violently knocked her chin against a counter. At the same time, during the fall, her tongue was somewhat severely bitten. There was profuse hæmorrhage, the wound was sutured and healed rapidly. Up to the age of eighteen she had perfect health. About this time, seeing unexpectedly a spider on her clothes, she experienced great fright, and her menstruation, till then regular, became suddenly and finally suppressed. Soon afterwards she began to suffer from violent attacks of headache, intermittent in character. At the age of twenty it was noticed that her intelligence was becoming deficient, and that she could not fix her attention. She showed at the same time physical inactivity, and general derangement of nutrition. About the same time she suffered from severe pain in a carious tooth in the upper jaw on the right side, which was extracted. Severe pains in the face on the same side then followed, which continued with more or less frequency up to the present time.

The patient then began to experience a general weakness in the joints of the upper and lower limbs. On one occasion, in consequence of a slight injury, a dislocation of the shoulder took place. Walking was often difficult on account of weakness of the knees and of the ankle joints. This con-

[1] We are indebted to M. P. Marie for the translation of this case, the existence of which was pointed out by M. A. Broca.

dition of the joints lasted several years. At the age of thirty the whole body became swollen. After some time this swelling went down in part, and is now found to be confined principally to the hands and face. At this time the tongue began to enlarge, and twelve months later changes in the lower jaw were observed; but it was not till a year later that the protusion of the lower jaw became sufficient to seriously inconvenience mastication and speech.

For five years she had had violent attacks of pain in the right ear with otorrhœa. Shortly afterwards an attack of epilepsy, rapidly followed by amaurosis and complete blindness (two years ago). These attacks have been repeated several times since.

On account of the constant and long-continued pressure of the hypertrophied tongue, the condyles of the lower jaw have been partially dislocated from the glenoid cavity, so that this bone is displaced downwards and forwards. It projects about an inch in front of the upper jaw, forming a considerable prominence with the chin. The face consequently appears lengthened, which, together with the hypertrophy of the lower jaw and foreshortening of the upper lip, gives the patient a peculiar and grotesque appearance, especially in profile. The tongue, which has undergone considerable changes in form and size, so as to be at times so swollen as to protrude from the mouth, was, when measured January 4th, 1857, three inches wide and three quarters of an inch thick. It presented deep indentations from the upper row of teeth. Mastication was very difficult and painful; speech also interfered with.

CASE XXXVII (Alibert).[1]

CASE XXXVIII (Sancerotte Noël). Marie, loc. cit.

[1] Described in the original paper by M. P. Marie in the 'Revue de Médecine,' to the translation of which the reader is referred for the notes. —TR.

BIBLIOGRAPHY OF ACROMEGALY.

Case I (P. Marie and S. Leite).—Souza-Leite's 'Thesis on Acromegaly.'
Case II (P. Marie and S. Leite).—Souza-Leite's 'Thesis on Acromegaly.'
Case III (P. Marie).—'Brain,' July, 1889. Additional notes by Souza-Leite, loc. cit.
Case IV (P. Marie).—' Nouv. Iconogr. Photograph,' 1888.
Case V. (P. Marie).—' Revue de Médecine,' 1886.
Case VI (P. Marie).—' Revue de Médecine,' 1886.
Cases VII and VIII. (P. Marie).—Souza-Leite, loc. cit.
Case IX. (Péchadre).—Souza-Leite, loc. cit.
Case X. (Farge).—Souza-Leite, loc. cit.
Case XI. (Flemming).—Souza-Leite, loc. cit.
Case XII. (Verstraeten).—Souza-Leite, loc. cit.
Case XIII. (Virchow).—Translated by Dr. A. A. Kanthack, 'The Illustrated Medical News,' vol. ii, p. 241.
Case XIV. (Freund).—Souza-Leite, loc. cit.
Case XV. (Roth).—Souza-Leite, loc. cit.
Case XVI. (Strümpnell).—' Munich Journal of Medicine.' Souza-Leite, loc. cit.
Case XVII. (Schultze).—Souza-Leite, loc. cit.
Case XVIII. (Schultze).—Souza-Leite, loc. cit.
Case XIX. (Alder).—Translated from the German by Dr. Bondaille: Souza-Leite, loc. cit.
Case XX. (Erb).—Souza-Leite, loc. cit.
Case XXI. (Erb).—Souza-Leite, loc. cit.
Case XXII. (Hadden and Ballance).—' Clinical Society's Transactions,' April, 1888.
Case XXIII. (Godlee). —' Clin. Soc. Trans.,' 1888. Souza-Leite, loc. cit.
Case XXIV. (Wilks).—Souza-Leite, loc. cit.

CASE XXV. (Tresilian).—Souza-Leite, loc. cit.
CASE XXVI. (Minkowski).—Souza-Leite, loc. cit.
CASE XXVII. (Wadsworth).—Souza-Leite, loc. cit.
CASE XXVIII. (Fritze and Klebs).—Souza-Leite, loc. cit.
CASE XXIX. (Ellinwood).—Souza-Leite, loc. cit.
CASE XXX. (Taruffi).—Souza-Leite, loc. cit.
CASE XXXI. (Brigidi).—Souza-Leite, loc. cit.
CASE XXXII. (Henrot).—Souza-Leite, loc. cit.
CASE XXXIII. (Lanceraux).—'Traité d'anatomie pathologique,' t. iii.
CASE XXXIV. (Lombroso).—Souza-Leite, loc. cit.
CASE XXXV. (Verga).—Souza-Leite, loc. cit.; autopsy by Marie (Nouvelle Icogn. de la Salpêtriére, 1889).
CASE XXXVI. (Chalk).—Souza-Leite, loc. cit.
CASE XXXVII. (Alibert).—Souza-Leite, loc. cit.
CASE XXXVIII. (Saucerotte-Noël).—Souza-Leite, loc. cit.

The above cases are given in Marie's and Souza-Leite's original essays. The order observed is that of the latter. Where the same case is described in both texts it is translated only in that of the former.—TR.

The following ADDITIONAL CASES have been collected from various sources, and are reprinted for the most part in the words of the respective authors.

CASE XXXIX. (Hutchinson), 'Archives of Surgery,' vol. i, p. 141. "A case of Acromegaly."—" The case of Mr. C. B——, of N——. His most prominent symptom has been a distressing and almost constant headache, which is mitigated only by taking food. The headache is always worse in the forenoon, beginning soon after breakfast, and steadily increasing till lunch. It never prevents his sleeping well, nor interferes with his eating. It is always present, more or less, when he wakes in the morning, and if by accident he is awake in the night he is always conscious of it. As a rule, however, it is never so bad in a recumbent position as when standing. He brings me a photograph taken ten years ago, which is in most singular contrast with the present appearance of his face. He had at that time a rather thin nose and finely marked features, whereas his nose has now become thick and tumid, and the skin of his face generally is almost like that of a leper. On the forehead it is thrown into thick, bossy folds, between which are deep wrinkles. At right angles with these folds, three or four furrows pass vertically upwards from the top of the forehead across his scalp. I never observed these vertical creases in any marked degree in any other person, but, no doubt, they are only exaggerations of what is normal. His skin is not only thickened, but it has become abnormally loose, and can be easily pinched or pushed into thick folds. The wrinkles on the forehead and the scalp are probably in part due to this looseness and thickening, and partly to the constant frowning which his headaches have caused. His ears are scarcely, if at all, involved in the general hypertrophy which the skin of the rest of his head has undergone. His age is 34. He has been married eight years, and his wife has borne seven children, of whom four are living and healthy. His grandfather had gout. His father, who came with him, is a florid, fine-looking man, in good health. In addition to the headaches and the tegumentary hypertrophy, Mr. B— shows a most remarkable condition of deformity of his fingers. These have all become twice their natural width, giving to his hand, naturally a large one, an enormous size. The increase in width is greatest near the end, and it gradually ceases towards the knuckles. It is not attended by any special enlargement of joints, nor is there any disease of the nails, excepting their increased width. The skin of his hands is thick, but it is not muddy, nor loose, like that of his face. On careful inquiry as to the order in which the symptoms had developed, it seemed certain that the enlargement of his fingers was noticed prior to the appearance of headaches. He consulted Dr. Garrod for the fingers more than two years ago, and at that time he believes he was not specially liable to headache. The iodide of potassium was prescribed with other remedies for the fingers. In connection with this I must here state that his father has an idiosyncrasy against the iodide, and has on two occasions experienced symptoms of poisoning from single small doses. On each occasion his face and scalp became much swollen. In my patient, his son, no obvious disagreement of the iodide was noticed, but it is from the time of its use, that the headaches date. About the time that the headaches began, Mr. B—, used to be much annoyed with singing in the left ear. This is still present to some extent, but has, as he described it, left the ear and become fixed in the middle of his head. It is not a constant symptom, but is usually present when the headache is bad. It occurred to me that it was possible that the case might be of the nature of myxœdema, but I could find no trace of general œdema, and no dulling of the intellect had been experienced. Mr. B— is a solicitor, and he never allows his headache to interfere with his attention to his profession. He is accustomed to sleep eight hours, to take a cold bath every morning, and to take either wine or beer twice in the day. Although his tongue is always furred and his urine always thick, yet he avers that attempts to restrict his diet have always had the effect of making him weak and miserable, without the slightest relief to his headaches. At one time he abstained for several months from stimulants, but found no benefit."

"As regards the measures of treatment which have been tried, it may be said that he has used almost all the ordinary forms of tonics, with purgatives, salines, and for a long time the bromide of potassium. Indian hemp he has pushed until it made him feel intoxicated and unfit for his occupation, but his headaches remained as before. There is but little to add that can help us to the discovery of the cause of the headaches. Mr. B— has never suffered from constipation, but he habitually loses blood by stool, without pain. When a boy he used to have pain, and probably had piles. He is not liable to cold feet. He never in his life had syphilis; and although both his father and grandfather have suffered from gout, he has never had it in any definite form. He once tried a complete rest in the South of Europe, and was away from business and his family for more than a month. Neither whilst away, nor after his return did he experience any alleviation of his headaches. He says that sometimes when out fishing, and enjoying a perfectly quiet day, he has been almost free from pain, and he notes definitely the influence of a liberal meal in giving him relief. Neither tea, coffee, nor stimulants are, he thinks, of any benefit. He used formerly to be a heavy smoker, but lately has been strictly moderate. In support of the suspicion that the iodide of potassium may have caused the face to swell, his father states that a year ago his face was much more swelled than it is now, especially under the eyes."

"Mr. B— called on me a second time in March, 1889. His aspect was decidedly less peculiar than it was two years ago. He looked less heavy and his features less thick, and the folds of his scalp were not so conspicuous. He was, he said, free from the discomfort in the use of his eyes which he had formerly complained of, and, on the whole, his headaches were less severe. The latter were, however, still very troublesome, and as formerly they usually came on in the forenoon. He had not been obliged to increase the size of his hats recently. The large size and heavy appearance of his lower jaw and lips were still very conspicuous, as also his giant fingers. The latter were large and flat, the flatness being especially marked at the joints, though it was not restricted to them. The terminal phalanges and the nails were widened also. The end of one forefinger around the last joint measured 2¾ inches, and the others were in proportion."

"I ascertained the following additional particulars: "His height without his boots is just six feet, his left foot he believes to have been for many years half an inch longer than the other. The foot measured from heel to end of great toe eleven inches. His toes were large and flat like his fingers, but not in similar degree."

"He had some difficulty in closing his hands owing to the clumsiness of his fingers; thus he could not make a fist. He could, however, grasp any large body, and he said that he could touch the keys of the pianoforte as well as ever. There was certainly some enlargement of the joint ends of his phalanges, but he had no arthritic pains. The fingers were chilly and showed some venous congestion. Their overgrowth ceased, or, almost so, at the knuckles. Certainly the skin and all the soft structures were involved in the overgrowth. The girth of his head was twenty-four inches, the height from ear to ear thirteen."

Cases XL, XLI, XLII (Hutchinson), 'Archives of Surgery,'

vol. ii, p. 297. "Three cases of Acromegaly illustrating the stage of Premonitory Symptoms." "I have had three cases of acromegaly under my observation during the last year. The first case was that of a lady, whom I had seen on a single occasion six or seven years ago without then making any diagnosis beyond that of rheumatism. This lady has recently come under the care of my friend Dr. Hughlings Jackson, by whom she was brought back to me as an example of Dr. Pierre Marie's malady. Such she now undoubtedly, is, her extremities being conspicuously enlarged, and her face lengthened. As she is now Dr. Hughlings Jackson's patient, and he will probably publish the case, I shall not give further details as to her present condition. It may be of interest, however, to transcribe from my note-book the particulars as regards her premonitory symptoms, as they were observed four years ago."

"Mrs. S—, æt. 32, suckling her child six months old, was sent to me November 15th, 1883, by Dr. Cæsar, of Tottenham. She complained of intense pain from the elbows downwards, and the right fingers were flexed into the palm. The other hand was sometimes affected in the same way. There was a sensation of "tingling numbness and intense burning" in the right hand. She had twice been threatened with rheumatism, but had then no joint affection. She had always been a chilly subject, and often thought that her blood did not circulate properly. She had been fairly well at the beginning of her lactation. The hand sometimes became "chalky-looking," white in colour, and she could not tell whether she had anything in her fingers or not. The first time she felt this was after sleep, when she thought her circulation had stopped. There was no difference between the two radials. She was often chilly and afterwards flushed."

CASE XLI. "My second case was that of Mr. A. G—. Mr. G— is aged 40, and has been married fourteen years. I first saw him in August, 1889, and again in June, 1890. He consulted me on account of defects of circulation in the extremities, with peeling patches in the ends of the fingers. On every occasion that I saw him I noticed the extraordinary size of his fingers and the length of his face, and several times pressed him as to whether he had not observed any increase in their growth recently. This at first he denied, but on a later occasion he admitted that the size of his gloves and shoes had been considerably increased. He tried the experiment with some old boots and found that he could not put them on. He brought me some photographs taken at intervals several years ago, and from them there could be no doubt that during the last six years the size of his face had conspicuously increased, especially in the lower part. It appeared, however, that he had always had a somewhat heavy lower jaw, and he said that his father, like himself, had large hands. The great size of his tongue was always a feature which attracted my attention. His countenance remarkably resembled that of Mr. C. B— (see 'Archives,' i, p. 141), the skin of his forehead being coarse, thick, and thrown into folds. The measurements of Mr. G—'s knuckles was 9¾ inches. The symptoms for which Mr. G— consulted me were the following:—He complained that he was liable to have the blood leave his hands and feet, and said that he felt weak and faint when it did so. He said that he was losing confidence in himself, and was becoming nervous. He was in the habit of rubbing the back of his neck violently when at business, in order to "revive his brain," as he said. In consequence probably of the constant disturbances in his circulation, the pulps of his fingers had become dry and cracked. He had not experienced any severe headache, but, like my patient Mr. B—, he had a longing for quiet and holiday. His tongue was habitually furred and white, although he had a good appetite. He suffered somewhat from piles."

"It will be seen that in these two cases the premonitory symptoms were not dissimilar. They consisted chiefly in derangements of circulation in the extremities, with lassitude."

CASE XLII. "My third case is that of a widow of a Welsh farmer, who was sent to me by Dr. W. Evans, of Anglesea."

"Mrs. H.— first came to me in May, 1887. She fancied she had gout, but I could find no proof of it, and prescribed a tonic. I did not then notice anything further than that her features and hands looked coarse and large. She suffered also from nasal polypus.

"Mrs. H.— came to me again three years later (September, 1890), and I then at once recognised that she was the subject of acromegaly. She said, when I suggested it, " Oh yes, I have been growing very much; my hands and feet are getting much bigger than they were." The enlargement was symmetrical in all parts, and appeared to shade off insensibly into the parts not involved. Her wrists and forearms were very large, and so was her whole head, without any bulgings. Her mouth and lips showed the condition most conspicuously. Her skin was very coarse and greasy. It was also constantly moist with perspiration. She was wearing large cloth shoes which laced, but which she could not nearly close. She considered herself rheumatic, but there was no stiffening of any joints. Her chief suffering was from facial neuralgia, which occurred in violent paroxysms, often keeping her awake at night. It was chiefly, but not exclusively, in the left side. She had not menstruated for three years or more. Her age was 46. Her legs, like her forearms, were very large. It is to be understood that the overgrowth was of skin and subcutaneous tissues quite as much as of bones. Although her face looked so large, and her lips especially, yet I could not on examination from within the mouth appreciate any thickening of the lower jaw. The left coronoid process appeared to be enlarged, but not much. The girth of her hands around the knuckles was nine inches and a half, and that of her index fingers three and a half."

"Mrs. H.— was a large, coarsely-built woman. She said that she had of late got weaker, but still attended to her household duties. The slightest exertion made her perspire."

CASE XLIII. " Acromegaly," ' Brit. Med. Journ.,' December 27th, 1890. "Dr. Joseph Redmond read and exhibited photograph, casts, &c., of a case of acromegaly.—A girl, æt. 19, was admitted into the Mater Misericordiæ Hospital on July 25th, 1890. Her illness commenced in December, 1889, during the epidemic of influenza. The first thing that attracted her attention was the condition of the hands and feet which were swollen, sore and tender to the touch. Her hands were large, fingers thickened and bulbous, and nails slightly convex. There was a marked increase in the size of the carpal ends of both radius and ulna. The backs of the hands were considerably swollen but did not pit on pressure. Her knees were enlarged. Below the knees the legs were uniformly enlarged. The ankles were widened and some effusion was present in the joints. The feet were larger and thicker than normal; the toes thickened and ends bulbous. There was œdema of the dorsum of the foot. Urine normal. The thyroid gland appeared to be absent, and no evidence of persistence of the thymus gland could be obtained. During her stay in the hospital she suffered occasionally from diarrhœa and pain in the back. Her pulse averaged 100; her temperature 101·6°, and her respirations 24."

CASE XLIV. 'Brit. Med. Journ., January 4th, 1890, vol. i, p. 19. "Acromegaly," &c.—"Mr. Silcock showed three cases of acromegaly, two being in women and one in a man. He drew special attention to the cases in which the disease, though easily demonstrable, was not so marked as in most of those which had hitherto been recorded, observing the condition was probably much more common than was supposed, because it had not been looked for."—Western district of the Metropolitan Counties Branch of the British Medical Association.

CASE XLV. Brit. Med. Journ., March 22nd, 1890, vol. i, p. 662. "Acromegaly" (Dr. Henry Waldo).—" W. R—, æt. 54, draper's assistant, was admitted on November 26th, 1889. Father and mother dead (cause unknown); no brothers or sisters; usual health good; no serious former illness. Six months ago patient found that his legs were getting weak, and that his knees were swollen. He also complained of great stiffness in his legs. He had difficulty in getting on his boots, and was obliged to procure a larger pair. Soon after this his hands and fingers began to get large. He found he could not do all his work, and had to give it up altogether a fortnight before admission, on account of weakness. Four days before admission he had a fit; he was noticed to be twitching all over; he then foamed at the mouth, and had a general convulsion; he slept for twenty-four hours after this. The patient is sallow and cachectic-looking, face thin, with skin shining and as if stretched over it. Zygomatic processes and orbital margins very prominent. Both hands appear to be too large for the man, they are enlarged in width and thickness, and they are clumsy-looking or paw-like. The fingers are all enlarged and thickened, all the structures of them appearing to participate in the enlargement, and the joints are not more affected than other parts. The thenar muscles are somewhat atrophied. He cannot close his hands, or indeed get his fingers beyond a right angle with the palm. His grasp is feeble, and he cannot button his clothes. The clavicles are both enlarged; the bones of the wrist appear thickened. On percussion of the upper region of the thorax, there is nothing corresponding to the triangular dulness of Erb to be made out. There is considerable enlargement of the knees, the left being the larger, and there is very little extra effusion, the enlargement being due to increase in the size of the patellæ and the ends of the long bones. The iliac crests are markedly thickened. There is no œdema anywhere. The veins of the arms and of the left knee are enlarged, and the finger and toe nails are more convex than usual. There is no enlargement of the head bones or lower jaw. The heart's impulse is in a line with the left nipple, and one inch and three fourths below it. No thrill. A systolic murmur is audible all over the front of the chest. Cough with slight mucous expectoration. Pleuritic friction with signs of fluid at base of right lung. The skin at top of sternum was very loose, and the laryngeal cartilages were voluminous. The left lobe of the thyroid gland could be felt, but not the right. Very little headache or other pain was complained of, and giddiness only when he sits up. The plantar reflexes were increased, the knee-jerks almost absent; pupils equal, act to light and accommodation, optic discs normal. His memory is very defective and has been so for six months; his mental processes are very slow, he is always drowsy, and his utterance drawling and monotonous. Latterly, he had delusions —thought people were going to kill him, and was suspicious about his food. He also developed symptoms of bulbar paralysis; he could not put out his tongue, he swallowed with the greatest difficulty (was fed per rectum), and he lost power of expectoration and of retaining urine (which was not albuminous). He had many attacks of retching and vomiting during the time he was in the Infirmary. Of the special senses, that of taste seemed lost. There was no numbness and the general cutaneous sensibility was normal. The temperature was subnormal throughout. He died on December 13th, 1889. The post-mortem appearances were an oval cavity in the brain substance, three quarters of an inch by one inch and a half at the posterior extremity of the right hemisphere; a cavity half an inch in diameter, three quarters of an inch from the posterior part of the second left temporo-sphenoidal convolution; a cavity in the anterior portion of each lateral cerebellar lobe, the left one inch by three quarters, the right a little smaller. The heart shows well marked aortic stenosis, all the cusps being calcareous. The left ventricular walls are hypertrophied. The right lung presented a large caseating mass of its lower lobe which was breaking down. The kidneys contained many small cavities in the cortex, varying in size from a pin's point up to a holly-berry; some contained pus and others a gummy material. The liver was nutmeggy. The left lobe of the thyroid gland was present, the right lobe was absent. There were no signs of a thymus gland. The pituitary gland was normal in size and appearance. No examination of the bones was made."

(This case is illustrated in the original text. The absence of enlargement of the pituitary gland was a peculiar feature. The case corresponds rather to the disease described by Marie as hypertrophic pulmonary osteo-arthropathy than to true acromegaly.—Tr.)

Case XLVI. (Dr. Robert Saundby), 'The Illustrated Medical News,' March 2nd, 1889, vol. ii, p. 195.—" John W—, æt. 37, stoker, was admitted into the General Hospital on August 21st, 1888, complaining of swelling of his hands, legs, and feet, and of pains in his wrists and knees.

Previous History.—He had suffered from bronchitis in the winter for four years, but had otherwise enjoyed good health. His work as a stoker obliged him to undergo great changes of temperature.

Family History.—His father had been rheumatic; father, mother, and three brothers were dead; one sister was alive and healthy. He was married, three healthy children were living, two were dead, one from convulsions, the other, cause not known.

History of Present Illness.—Fourteen weeks before admission he noticed that he had difficulty in pulling on his boots. The swelling began on the dorsum of the foot. A fortnight later he had pains in his ankles and knees, and soon after in the wrists, while his knees swelled and grew stiff. At the end of three weeks swelling began on the backs of his hands, while pains in his wrists obliged him to give up work. The pain was aching in character, and was brought on by moving the joint. His limbs became weak, while his feet grew so large that he could not get his boots on. At this time he began to be thirsty, so that he took water to his bedside at night. His feet attained their present size in about three weeks, while the hands grew more rapidly, reaching their present condition in about a week.

Present condition.—On admission he no longer suffered from thirst. He was a fairly developed, poorly-nourished man, with a sallow complexion. Neither he nor his wife had noticed any alteration in his face. Temp. 98·5°; resp. 20; pulse 90. The most striking thing about him was the great enlargement of his hands and forearms, of which the bones as well as the soft parts were greatly hypertrophied. The left radius was twice the natural thickness, and the ulna was thickened at its lower extremity. The right radius and ulna were even larger. The right hand was larger than the left. The fingers were thickened and their ends bulbous, those of the right hand being larger. The nails were more convex than normal.

The skin was not œdematous or thickened, but its wrinkles were wider than normal, and it was discoloured with patches of yellow pigment. The hair on the hands was normal.

The veins of the forearm were greatly swollen, and the radial artery was much enlarged; the pulse was very large, full, and soft.

The knees were enlarged, the right being the bigger; the enlargement being in part due to synovial effusion, in part to enlargement of the patella and the head of the tibia and fibula. The veins on the knees were large and full. The legs below the knees were uniformly enlarged down to the ankles, the right being the larger; there was increased local swelling about the ankle-joints, due to synovial effusion. The feet were increased in breadth and thickness, calling to mind the feet in the early stages of elephantiasis. The toes were thickened. There was a little œdema of the legs and feet. There was no abnormality about the nails, skin, or hair of the lower extremities. The movements of all the affected joints were diminished. He could not close his right hand completely.

The grasp of both hands was weakened; his arms were soon tired, and he felt clumsy with his hands, though he could not button his clothes, &c. Walking caused pain behind the knees and in the ankles, the ribs, clavicles, and iliac crests felt thickened. The ears were large, and their fibro-cartilages thickened. The cartilages of the larynx seemed normal.

The skin on the trachea was quite loose, and no thyroid body could be felt. His face was thin, thinner than it used to be.

The zygoma and malar bones were prominent. The nose was large; the fibro-cartilages were probably thickened. The chin was long and pointed, but the bone did not feel thickened, and compared with an old photograph appeared to be natural.

The sutures of the skull were not particularly well marked, with the exception of the sagittal suture. There was no special development of the various protuberances.

His mental condition on admission seemed quite normal, but before he left he seemed to be at times rather rambling and childish.

There was no loss of memory, headache, affection of special senses, or vomiting. He complained of numbness in his hands, but general cutaneous sensibility was normal.

The muscles had wasted, and there was diminution of muscular power. Most of the muscles showed local contractions when lightly percussed. The patellar and bicipital reflexes were absent. The plantar and other superficial reflexes were exaggerated.

The pupils reacted to light and accommodation.

The tongue was not enlarged; his teeth were defective, alimentary system otherwise normal.

Heart not enlarged; sounds pure but feeble; pulse full, regular, compressible, 90 to 96. Breath-sounds harsh posteriorly; some cough, with expectoration of thick muco-pus. Urine 1020, alkaline; no albumen or sugar.

The temperature was taken throughout, and for the most part ranged between 99° and 100°, occasionally reaching 101° F. There was no sweating. He suffered a good deal from bronchial catarrh, and was evidently getting weaker.

On October 18th he rather abruptly left the hospital. About ten days later I heard of his death, and through the kindness of his medical attendant, Mr. C. W. Biden, a reluctant consent to an examination was obtained. The necropsy was performed by Dr. Crooke, in the presence of Mr. Biden and myself, at the patient's own house, and under unfavorable circumstances. The head, thorax, and abdomen were alone examined.

Date of death, October 28th, 8 a.m.; date of examination, October 29th, 4.45 p.m. Length of body from heel to vertex, 5 feet $7\frac{1}{2}$ inches. Rigor mortis had passed off; hypostatic congestion was well marked.

Head: Scalp thick; skull-cap dense and heavy, sutures massive; thickness of frontal bone, maximum quarter of an inch, minimum three sixteenths of an inch; occipital bones, maximum quarter of an inch, minimum quarter of an inch. Pia mater congested throughout; brain substance uniformly congested, and vessels full of blood; pituitary body normal in size and appearance.

Thorax: Thyroid body atrophied. Heart enlarged, full of fluid blood and recent clots; walls flabby; right ventricle quarter of an inch thick; left ventricle four sixteenths of an inch thick; valves normal.

Left lung engorged, œdematous, and slaty; in the upper part of the lower lobe, posteriorly, was a greyish rounded mass as large as an orange, extending backwards to the root of the lung; glands at root black and pigmented, but not enlarged.

Right lung: margin emphysematous; at the lower border of lowest lobe, laterally, was a projecting growth the size of a walnut, covered by smooth pleura; the root of the lung was in the same condition as the left.

Abdomen: Liver nutmeggy, fatty, and enlarged; no new growth.

Spleen: Small, normal.

Kidneys: The two kidneys were united at the lower end, so as to form a very good specimen of a horseshoe kidney, but appeared normal.

There were no other growths to be felt, nor any enlarged glands. The clavicles were distinctly enlarged. A portion of the left brachial plexus was removed for microscopic examination.

Microscopic examination. — The thyroid body showed thickening of its stroma; only a relatively small number of alveoli contained colloid matter; numbers were full of a granulation tissue, while in large tracts the alveolar structure was disappearing in a lowly organised nucleated connective tissue. The pituitary body appeared quite normal.

The growth in the lung was a spindle-celled sarcoma growing from the wall of a branch of the pulmonary artery. In its neighbourhood the lung-tissue was infiltrated by caseating pneumonia. The nerves showed some cloudiness of certain bundles, and on longitudinal section, in the osmic acid preparations, the neurin was coagulated in little lumps outside and along the walls of the nerve-tubules. There was no increase of nuclei or any distinct appearances of either parenchymatous or interstitial neuritis."

CASE XLVII. "A case of Acromegaly."[1]—(Dr. A. A. Kanthack), 'Brit. Med. Journ.,' July 25th, 1891.—" Hoshnaki, a Ghirth by caste, aged 16 to 18 years, comes from the Rangra Valley. He has a slight cystic goître, more marked on the left side—an exceedingly common affection in this valley—is five feet four inches high and of only moderate muscular development. The most striking part of his body is his left

[1] Measurements of parts given.—TR.

foot, which is much enlarged. The second toe is immensely hypertrophied, all constituents, bones, and subcutaneous tissue being much increased, but chiefly the latter. The third or middle toe is likewise considerably enlarged, and the other toes are all thickened. The right foot is just commencing to be affected, all the toes becoming thickened, and thus looking "stumpy." His left hand is apparently normal, but his right hand is beginning to show early signs, all the fingers becoming thickened, due to increase of the subcutaneous tissue, and already recall to one's mind the condition of the fingers described by Virchow as "tatzenartig" (paw-like). The lower part of his face is broadening out, and the angles of the lower jaw becoming prominent. He himself has noticed these changes. Otherwise his limbs are perfectly natural. It is very interesting that this patient is suffering from goître, as by several authors absence of the thyroid gland has been considered of etioligical importance. Here, on the other hand, we have a hypertrophy of the thyroid. This condition of things has, however, also been noticed in some other recorded cases of acromegaly. A hereditary taint could not be traced, as was the case in Virchow's case—Mennig.[1]"

"Unfortunately, I had no means of taking head measurements with any degree of accuracy, nor did I find an opportunity of examining his urine for sugar. In several cases lately, it has been observed that sugar at times in apparently large quantities exists in the urine; the latter, however, remaining always of a dark colour. Thus, when I had an opportunity a few months back to examine the urine of "Goliath" (case Westphalen, of Virchow), Fehling's solution was copiously reduced."

"I think, from the whole appearance of the youth, there can be no doubt that this is a genuine case of acromegaly. The hand measurements do not seem astonishing at first sight, but there is (1) a marked disproportionate difference between the two hands in favour of the right hand ; (2) the

[1] It seems almost a pity that this case should be recorded as acromegaly. The enormous hypertrophy of the second toe of the left foot is probably an example of congenital overgrowth, and the other conditions appear to have been in too early a stage to permit of any confident diagnosis.—Tr.

thickness of the fingers and toes become especially apparent on comparing them with those of other natives, who have notably thin fingers and toes. He cannot remember when the changes began, but states that ever since he can recollect, the left foot had been affected and gradually increasing, having now, however, become stationary. He does not acknowledge any changes in his right foot, though he confesses that he has noticed the difference between the two hands, which has lately become more marked. For the photograph I am indebted to my friend, Surgeon-Major A. Barclay, and I take this opportunity of thanking him for his kind assistance."

CASE XLVIII. " A case of Acromegaly "[1] (Dr. Robert Ruttle), ' Brit. Med. Journ.,' March 28th, 1891.—" The first of the accompanying photographs is that of the patient just before her marriage in 1876, at that time she was 24 years of age. The second, taken recently in the same position, affords a striking contrast.

"I saw the patient in May, 1879. She was not then or since until October 11th, 1890, a patient of mine; but at that casual meeting her face, hands, and feet at once attracted my attention. I thought of cretinism, but the patient was clear, cool, intelligent, though certainly in a peculiar condition, and she let fall the remark that she had been under Dr. Popjoy for a considerable time. She is now thirty-eight, has been married fourteen years, has had no family. She has been accustomed to heavy farm service, describes herself as ' strong and healthy, with cheeks like roses,' up to the time of first symptoms. She had no severe illness nor any irregularity, except that her menstrual periodicity varied slightly, but without pain or any other inconvenience; she menstruated once after marriage, never since then. Soon after she began to experience severe pains about vertex, back of head, and in the cervical region, which increased gradually in intensity; she also complained of a feeling of ' numbness ' all over her body; the headache became ' dreadful,' and she then noticed that her hands and feet were ' swelling,' because No. 7 gloves no

[1] Shown at the Blackburn and District Medical Society.

longer fitted her hands, and 5 in boots also became too small. This enlargement went on gradually. When she was twelve months married she was suddenly seized in the mill with an attack of vertigo, followed by total loss of consciousness. She was taken to a friend's house, and from there removed home with difficulty, kept her bed for nearly three weeks, hanging apparently between life and death. Her convalescence was slow, alternating with relapses of agonising pain, accompanied by vertigo and loss of consciousness. These attacks seemed of late years to have somewhat decreased in severity. Treatment always was powerless to relieve her, rest and quiet alone doing any good ; she appeared better when idle in the country. At the time of the severe explosion her extremities rapidly enlarged, so much so that her left finger burst her wedding ring, a moderately stout one. She also began to experience failure of sight at this time, which has gone on to total blindness in the right eye from atrophy of the optic nerve. In the left eye the temporal half of the field of vision is gone, and the activity of the upper limit at the nasal side slightly encroached on ; the hemianopia is very clearly defined ; refraction is emmetropic, the pupils always somewhat dilated ; the optic axes are divergent.

"Heart, liver, kidneys, &c., appear sound, no albumen or sugar present ; occasionally a slight cloud of phosphates in urine. Digestion fairly good, bowels inclined to constipation ; she eats little, often not till evening. She can sleep well, but wakens often during the night. I might add that her pain now generally comes on at night and wakens her ; but sleep and rest afford relief, stimulants rather increase pain. She does not use tea. Coffee gives some slight relief ; coffee and cocoa are her usual beverages."

"Her face is pale, with skin smooth and few wrinkles, but not very marked, slight puffiness under lower lids certainly varies, while eyes are somewhat divergent and lids large ; the cartilages are enlarged, nose long and very large ; both lids hypertrophied, especially the lower one ; the lower jaw and chin enlarged, prominent, prognathous ; teeth separated, tongue large ; the face does not appear at all broadened nor the frontal sinuses enlarged. Her head measures twenty-three and a quarter inches in circumference round the brows.

"She has become somewhat stooped, and complains of sciatic pains, especially on the left side, which compel her to walk lamely. Her left hand measures eight and a quarter inches in length, and is eight and three quarter inches round the knuckles; the second joint of the left forefinger is three and a quarter inches in circumference; no women's gloves or boots fit her; she wears 8½ boots. She can now close her hand and make a fist. Both radius and ulna are enlarged, so are the malleoli and heads of tibiæ. The vertebræ appear normal in size, but the sternum is certainly larger than normal; the pelvis also is certainly larger; the ilium feels most massive. She says she 'hardly had any hips before.' She has had dyspareunia for two years; the vaginal orifice appears contracted spasmodically, though it allowed the passage of the fingers, when the os tincæ and the canal of the vagina appeared normal. She had slight eczema of external auditory meatus, but no appreciable enlargement of the ears. She complains of great sensitiveness over the lower cervical and first dorsal vertebræ. A playful slap on the shoulders has often produced unconsciousness. She cannot bear the least noise; a sudden noise as of a door slamming will often act like a blow."

"I can elicit no family history of gout, syphilis, or tubercle. Her father is still living; her mother died at sixty-eight; brothers and sisters are healthy. Her intelligence is clear, and quite above the average of women in the lower station of midland rural life. She presents no deterioration of faculty save that of sight. The thyroid is enlarged, and I believe the thymus also. I may here be permitted to point out the most noteworthy features of this particular case:

1. The stoppage of catamenia.
2. The enlargement of extremities without pain.
3. Pain in the head (vertex, without enlargement).
4. The eye symptoms, R. atrophy, L. temporal hemianopsia.
5. The abnormal sensitiveness to sound.
6. The tenderness and sensitiveness over her cervical and upper dorsal region.

I think that we have here enough to point to derangement

of a neurotrophic character, of the precise characters of which we are as yet, I fear, ignorant. The atrophy and hemianopsia certainly favour Dr. Marie's views respecting an enlarged pituitary body."

"The only drug which I have found to give her any relief has been exalgine, in doses of three grains repeated every eighth hour. This almost completely relieves the migraine, and has given her great ease from the sciatic pain."

INDEX.

ACROMEGALY, age at onset of, 13, 59
— bibliography of, 161
— diagnosis of, 67
— etiology of, 13
— general health in, 13
— pathology of, 61
— summary of, by Marie, 27
— symptoms of, 37, 60
ALBUMINURIA, 53, 95
APPETITE, increased, 85

BIBLIOGRAPHY, 161

CARTILAGES, 17, 18, 43
CASES, 3, 81
— additional, 101
CEPHALALGIA, 4, 10, 55, 85, 110, 123, 143, 159, 162, 177
CLAVICLES, changes in, 10
CRANIUM, 11, 46

DIAGNOSIS, 66

ELECTRICAL CHANGES, 116
ETIOLOGY, 13, 34
EXOPHTHALMOS, 19, 42, 56, 140

FACE, changes in, 11, 32, 41, 46, 84, 98, 129, 137, 138, 140, 154, 156, 178
FEET, enlargement of, 4, 112, 127, 135, 142, 144, 167, 170
FOREARMS, changes in, 17

GENERATIVE ORGANS, 50, 115
GIDDINESS, 103
GIGANTISM, diagnosis from acromegaly, 69
GLANDS, changes in, 19

HAIR, 88, 146

HANDS, changes in, 7, 16, 32, 37, 111, 130, 144, 151, 167, 170
— in osteo-arthropathy, 73
HEARING, 12, 57, 139, 141, 160
HEART, 51, 131

JOINTS, 5, 12, 51
— in osteo-arthropathy, 77

LARYNX, 40, 85, 140
LEGS, changes in, 17
LIPS, changes in, 43
LOWER JAW, changes in, 45

MENSTRUATION, 1, 55, 102, 119, 125, 133, 138
MENTAL SYMPTOMS, 57
MOLLUSCUM FIBROSUM in acromegaly, 54, 91
MUSCLES, 51, 118
MYXŒDEMA, diagnosis from acromegaly, 68

NAILS, changes in, 10, 113, 116
NERVES, disease of, 20
NEURO-RETINITIS, 62
NOSE, 19, 84, 140, 142

OSTEITIS DEFORMANS, cases of, 23, 24
— — diagnosis from acromegaly, 25
OSTEO-ARTHROPATHY, diagnosis from acromegaly, 70

PAINS, in limbs, &c., in acromegaly, 55
PATHOLOGY, 61, 148, 155, 174
PELVIS, changes in, 50, 127
PERSPIRATION, 53, 103
PITUITARY GLAND, 20, 62, 155
PROGNOSIS, 59

INDEX.

RAYNAUD'S MALADY, diagnosis from acromegaly, 78
RHEUMATISM, chronic, diagnosis from acromegaly, 69

SENSATION, changes of, 54
SIGHT, 12, 56, 136
SINUSES, 43
SKIN, 13, 54, 98, 113, 114, 123, 131, 145, 156, 163
SMELL, 57
SPINE, changes in, 47, 71, 127, 141, 145, 151, 153
SYMPATHETIC, changes in, 21, 63
SYMPTOMS, 37

TASTE, 57
TEETH, 99, 117, 137, 151, 156
THIRST, excessive, 9, 172
THORAX, 33, 47, 97, 114, 127, 144, 147
THYMUS, 63
THYROID GLAND, 5, 19, 99, 134, 136, 140
TONGUE, hypertrophy of, 12, 19, 104, 140
TREATMENT, 80

UNILATERAL HYPERTROPHY, diagnosis from acromegaly, 22, 78
URINE, excess of, 13, 53, 91, 141

VARICOSE VEINS, 13, 98, 131

PRINTED BY ADLARD AND SON, BARTHOLOMEW CLOSE.

REPORT

PRESENTED TO THE

THIRTY-THIRD ANNUAL MEETING

OF THE

NEW SYDENHAM SOCIETY

HELD AT BOURNEMOUTH,

JULY 31ST, 1891.

WITH

BALANCE SHEET FOR 1890, LIST OF OFFICERS FOR 1891-92,

AND LIST OF PUBLISHED WORKS.

———◆———

AGENT AND DEPÔT FOR THE PUBLICATIONS,

H. K. LEWIS, 136, GOWER STREET, LONDON.

OFFICERS FOR 1891-92.

President.
Sir JAMES PAGET, Bart., F.R.S., LL.D., D.C.L.

Vice-Presidents.

*JAMES ANDREW, M.D.
R. L. BOWLES, M.D. (Folkestone).
THOMAS BRYANT, Esq.
THOMAS BUZZARD, M.D.
Sir ANDREW CLARK, Bart., F.R.S., LL.D.
J. LANGDON DOWN, M.D.
C. J. HARE, M.D.
W. M. GRAILY HEWITT, M.D.
CONSTANTINE HOLMAN, M.D. (Reigate).
D. J. LEECH, M.D. (Manchester).

Sir JOSEPH LISTER, Bart., F.R.S.
*Sir DOUGLAS MACLAGAN, M.D. (Edinburgh).
Sir WILLIAM ROBERTS, M.D.
Sir JAMES SAWYER, M.D. (Birmingham).
*THOMAS SCATTERGOOD, Esq. (Leeds).
SEPTIMUS W. SIBLEY, Esq.
*J. R. THOMSON, M.D. (Bournemouth).
HERMANN WEBER, M.D.

Council.

THOMAS BARLOW, M.D.
MARCUS BECK, Esq.
*A. A. BOWLBY, Esq.
*BYROM BRAMWELL, M.D. (Edinburgh).
J. W. BYERS, M.D. (Belfast).
ALFRED H. CARTER, M.D. (Birmingham).
W. B. CHEADLE, M.D.
W. WATSON CHEYNE, Esq.
W. CHOLMELEY, M.D.
W. S. CHURCH, M.D.
*H. RADCLIFFE CROCKER, M.D.
F. R. CRUISE, M.D. (Dublin).
W. CLEMENT DANIEL, M.D. (Epsom).
J. N. C. DAVIES-COLLEY, Esq.
JULIUS DRESCHFELD, M.D. (Manchester).
*A. E. W. FOX, M.D. (Bath).

*T. R. GLYNN, M.D. (Liverpool).
J. F. GOODHART, M.D.
G. E. HERMAN, M.D.
JAMIESON B. HURRY, M.D. (Reading).
W. HAMERTON JALLAND, Esq. (York).
W. ALLAN JAMIESON, M.D. (Edinburgh).
*P. W. LATHAM, M.D. (Cambridge).
H. CRIPPS LAWRENCE, Esq.
W. WARD LEADAM, M.D.
MONTAGU LUBBOCK, M.D.
STEPHEN MACKENZIE, M.D.
*THOMAS OLIVER, M.D. (Newcastle).
T. PICKERING PICK, Esq.
G. H. SAVAGE, M.D.
*F. CHARLEWOOD TURNER, M.D.
*F. P. WEAVER, M.D. (Frodsham).

Treasurer.
W. SEDGWICK SAUNDERS, M.D., F.S.A., 13, Queen Street, Cheapside, E.C.

Auditors.
E. CLAPTON, M.D. | JOHN CROFT, Esq.
F. M. CORNER, Esq.

Hon. Secretary.
JONATHAN HUTCHINSON, Esq., F.R.S., LL.D., 15, Cavendish Square, W.

Those marked with an Asterisk were not in office last year.

REPORT

PRESENTED TO THE THIRTY-THIRD ANNUAL MEETING OF THE NEW SYDENHAM SOCIETY, HELD AT BOURNEMOUTH, JULY 31ST, 1891.

During the past year (1890) the following four works have been issued :—

> Flügge's Treatise on Micro-parasites.
> The third and concluding volume of Cohnheim's Lectures on Pathology.
> The seventeenth part of the Lexicon of Medical Terms.
> The eighth Fasciculus of the Atlas of Pathology (comprising Diseases of the Brain and Spinal Cord).

With the object of hastening the completion of the Lexicon of Medical Terms, it has been decided to associate a third Editor with Mr. Power and Dr. Sedgwick. This arrangement will somewhat increase the cost of an undertaking which has already proved very expensive, but it will, it is trusted, secure its early completion, an end which is felt to be of paramount importance.

The Society's accounts for the year have been audited, as usual, and the Balance Sheet is annexed. It will be seen that the sum carried over was smaller than in some former years, and less than is desirable for the prosperity of the Society. It is satisfactory to know, however, that there has been no falling off in the list of members, but rather the reverse. The Council believes that it would be easy, by additional exertion on the part of Local Secretaries and individual members, to place the Society's funds in a much better position, and thus enable it to issue additional volumes, and at the same time to keep in the Treasurer's hands a better balance.

The Council has during the year had many works under consideration, with a view to publication, and has adopted several.

It is probable that the following will constitute the series for the current year (1891) :—

    I. Ewald's Lectures on Disorders of Digestion. This work, translated by Dr. Saundby, will be ready for issue in about a month.

    II. Acromegaly. Dr. Pierre Marie's Original Essays, and Dr. Souza Leite's more recent Thesis. With Illustrations.

    III. The eighteenth part of the Lexicon of Medical Terms.

    IV. A volume of *Clinical Lectures and Essays from German sources*. This volume will contain, with others, the following:—1. Billroth on the Mutual Relations of Living Animal and Vegetable Cells. 2. Von Ziemssen on Neurasthenia. 3. Von Ziemssen on the Etiology of Tuberculosis. 4. Salzer on the Process of Healing-in of Foreign Bodies, with Lithographs.

V. A volume of *Dermatological Papers and Lectures.* This volume will contain, with others :—

Dr. Prince Morrow's work on Drug Eruptions (edited, with notes, by Dr. Colcott Fox).

Selections from the Writings of Dr. Unna.

Dr. White on Keratosis Follicularis.

Dr. Ludwig Berger on Pellagra.

Amongst the works which the Society has in hand, but which will not form part of the current year's series, are the following :—

Additional Fasciculi of the Atlas of Pathology.

A volume of Selected Monographs on Gynecological Subjects.

Selected Lectures from Jaccoud's Clinical Medicine.

Selections from the Works of Prof. Fournier.

THE NEW SYDENHAM SOCIETY.—BALANCE SHEET FOR 1890.

Receipts.

	£	s.	d.	£	s.	d.
Balance in hand brought forward from last account				354	2	4
Subscriptions.. 11 for 1869 to 1884	11	11	0			
" 9 " 1885	9	9	0			
" 22 " 1886	23	2	0			
" 123 " 1887	129	3	0			
" 238 " 1888	249	18	0			
" 516 " 1889	541	16	0			
" 1394 " 1890	1463	14	0			
" 16 " 1891	16	16	0			
Say, 2329 Subscriptions	2445	9	0			
Back Volumes	185	13	6			
Repayment of postage, &c., charged in Agent's Disbursement Account	42	18	9			
	2674	1	3			
Less deductions by Local Secretaries	27	7	7			
				2646	13	8
				£3000	16	0
To Balance brought down				125	16	11

W. SEDGWICK SAUNDERS,
Treasurer.

Expenditure.

	£	s.	d.	£	s.	d.
Artists, Editors and Translators, Printers, Paper, and Bookbinders				2238	17	10
Expenses of Management:—						
Agent's Salary and Percentage	273	10	3			
Disbursements (chiefly carriage of books)	103	19	0			
Advertisements	136	16	0			
Fire Insurance of Stock	14	10	0			
Treasurer's Expenses	2	15	0			
Assistant Secretary for 1890	54	11	0			
				586	1	3
				2874	19	1
Balance in hand				125	16	11
				£3000	16	0

Examined, compared with the vouchers, and found correct, the balance on 31st Dec, 1890, being £125 16s. 11d., at an audit held this 9th day of July, 1891.

EDWARD CLAPTON
F. M. CORNER } *Auditors.*
JOHN CROFT

LONDON, July, 1891.

CLASSIFIED LIST

OF THE

SOCIETY'S PUBLICATIONS.

Medicine.

MICRO-ORGANISMS, WITH SPECIAL REFERENCE TO THE ETIOLOGY OF THE INFECTIOUS DISEASES. By Dr. C. FLÜGGE, O. O. Professor and Director of the Hygienic Institute at Göttingen. Translated by W. WATSON CHEYNE, M.B., Surgeon to King's College Hospital. With 144 Drawings.

"This volume forms an important addition to English medical literature. Flügge's book being justly considered one of the best standard text-books."—*British Medical Journal.*

"This translation is a most important addition to the English literature concerning Bacteria, and well deserves a place beside the most important volumes hitherto issued by the New Sydenham Society. . . . . . The work is most valuable, and we can cordially recommend it to all who take an interest in Micro-organisms. To Mr. Watson Cheyne's work as a translator, too high praise cannot be given. We have rarely met with a translation which could be read with as much pleasure, and in which the language and style were as good."
—*Dublin Medical Journal*, Sept. 1890.

LECTURES ON GENERAL PATHOLOGY. 3 Vols. By JULIUS COHNHEIM. Translated from Second German Edition by ALEXANDER B. MCKEE, M.B., Dublin.

"The publication of the work wherein the late Prof. Cohnheim may be said to have gathered the harvest of his remarkable scientific career is an event regarding which the New Sydenham Society deserves great credit. . . . . The influence of Cohnheim's teaching has shown itself in many recent works on Pathology, but hitherto the English-reading student has had to be content with the descriptions of his work at second-hand. He will now have the opportunity of studying the words of the teacher himself, and he cannot fail to be impressed by the depth of knowledge and clear thinking that are evinced on every page. For this boon he has to thank the New Sydenham Society, and especially the able translator, Dr. McKee, who has succeeded admirably in the by no means easy task of rendering such a work as this into another language, without impairing the freshness or terseness of the original."—*Lancet*, Sept. 21, 1889.

"The New Sydenham Society has done a good work in making Cohnheim's famous lectures accessible to the English reader."—*British Medical Journal.*

"Dr. McKee has discharged a difficult duty most ably, and conferred a boon on British students of pathology."—*London Medical Recorder.*

"The excellence of the author's work is retained by the care and ability with which the Lectures are done into English by Dr. McKee, and the two volumes form a useful and welcome addition to the list of valuable books provided for the profession by the New Sydenham Society."—*Medical Press.*

LECTURES ON CHILDREN'S DISEASES. 2 Vols. By Dr. C. HENOCH. Translated from the Fourth Edition (1889) by JOHN THOMSON, M.B., F.R.C.P. Edinb.

"The clinical types are depicted with the hand of a master, and the remarks upon etiology and treatment are exhaustive and precise. It is long since we have read any book with more pleasure and profit than we have experienced in perusing Prof. Henoch's work in its English dress." — *Glasgow Med. Journal,* Aug. 1889.

"It is an exceedingly valuable work, reflecting as it does the very best clinical opinion in Germany. Useful hints may be gathered on almost every page; few authorities are quoted, the author mainly relying on his own varied clinical experience, which has now extended over forty-five years. Dr. Thomson has done his work as a translator well, and has succeeded in producing a readable English version of a most valuable text-book."—*British Medical Journal.*

RECENT ESSAYS BY VARIOUS AUTHORS ON BACTERIA IN RELATION TO DISEASE. Selected and Edited by W. WATSON CHEYNE, M.B., F.R.C.S.

"This is a valuable collection of some of the most important papers on Bacteriology which have appeared in Germany, including Koch's papers on the investigation of Pathogenic Organisms, the Etiology of Tuberculosis, and the Etiology of Cholera; Frieländer's paper on the Micrococci of Pneumonia; and others on Leprosy, Enteric Fever, Glanders, &c., by well-known bacteriologists. The work of translation has been uniformly well done, and has been distributed among a large number of collaborators. The volume has had the great advantage of being edited by Mr. Watson Cheyne, and will be highly prized by all members of the Society as a most useful and interesting addition to their book-shelves."—*Birm. Med. Review*, July, 1886.

GEOGRAPHICAL AND HISTORICAL PATHOLOGY. Vols. I., II., and III. By Dr. AUG. HIRSCH. Translated from Second Edition by CHARLES CREIGHTON, M.D.

Vol. I. "The Council of the New Sydenham Society has done the Profession in England a right good service in having brought the great work of Professor Hirsch, of Berlin, within the reach of English readers; and it has been particularly fortunate in having enlisted the services of Dr. Charles Creighton in the very onerous task of translating so large a volume. The third chapter, dealing with Sweating Sickness, is perhaps the most interesting in the whole volume, and contains a vast amount of information which the reader will search for in vain elsewhere."—*Medical Times and Gazette*, May 10, 1884.

"The book is indeed a marvel of industry and erudition, and one which ought to be consulted by every writer on Medicine; no summary will, however, suffice to indicate the wealth of material so laboriously collected and so skilfully arranged, and our readers must turn to the volume itself, which will well repay perusal."—*Lancet*, July 12, 1884.

Vol. II. "It is a deep mine of facts and information combined, and judiciously arranged by the learned author; and Dr. Creighton has admirably performed his part in presenting it in an attractive English dress."—*Dublin Medical Journal*, Sept., 1884.

ON THE TEMPERATURE IN DISEASE: A MANUAL OF MEDICAL THERMOMETRY. By Dr. C. A. WUNDERLICH. (Leipzig). Translated by Dr. BATHURST WOODMAN. With forty Woodcuts and seven Lithographs.

"It is a work of reference absolutely necessary for all who would keep themselves abreast of the day in relation to so important a matter as corporeal temperature."—*Edin. Med. Journ.*

LECTURES ON CLINICAL MEDICINE, delivered at the Hotel Dieu, Paris. By Professor TROUSSEAU. Five Volumes. Vol. 1, translated, with notes and appendices, by the late Dr. BAZIRE. Vols. 2 to 5, translated from the third edition, revised and enlarged, by Sir JOHN ROSE CORMACK.

"We are indebted to the New Sydenham Society for this rich contribution to our medical literature. . . . . Trousseau is an author to be read rather than reviewed. He can only be criticised worthily at the bedside. . . . . . We commend this great physician's work to the study of every reader."—*Lancet*.

LATHAM'S COLLECTED WORKS. 2 vols. Edited by Dr. ROBERT MARTIN. With Memoir of LATHAM by Sir THOMAS WATSON.

LOCAL ASPHYXIA AND SYMMETRICAL GANGRENE OF THE EXTREMITIES. By MAURICE RAYNAUD. Translated by Dr. THOMAS BARLOW.

ON THE NATURE OF MALARIA. By Professors EDWIN KLEBS and C. TOMMASI-CRUDELI; and ALTERATIONS IN THE RED GLOBULES IN MALARIA INFECTION; and ON THE ORIGIN OF MELANÆMIA. By Professor ETTORE MARCHIAFAVA and Dr. A. CELLI. Translated by Dr. E. DRUMMOND, of Rome.

CLINICAL LECTURES ON MEDICINE AND SURGERY. Translated from the German, and selected from Professor Volkmann's Series. Two Volumes.

MEMOIRS ON DIPHTHERIA; containing Memoirs by Bretonneau, Trousseau, Daviot, Guersant, Bouchet, Empis, &c. Selected and Translated by Dr. R. H. SEMPLE.

RADICKE'S PAPERS ON THE APPLICATION OF STATISTICS TO MEDICAL INQUIRIES. Translated by Dr. BOND.

LECTURES ON PHTHISIS. By Professor NIEMEYER. Translated by Professor BAUMLER.

THE COLLECTED WORKS OF DR. ADDISON. Edited, with Introductory Prefaces to several of the Papers, by Dr. WILKS and Dr. DALDY. Portrait, and numerous Lithographs.

A GUIDE TO THE QUALITATIVE AND QUANTITATIVE ANALYSIS OF THE URINE. By Dr. C. NEUBAUER and Dr. J. VOGEL. Fourth edition, considerably enlarged. Translated by WILLIAM O. MARKHAM, F.R.C.P.L. With four Lithographs, and numerous Woodcuts.

"The New Sydenham Society have conferred a benefit, not only on their own subscribers, but on the whole profession in this country, by publishing the work of Drs. Neubauer and Vogel."—*Medical Times and Gazette.*

MEMOIRS ON ABDOMINAL TUMOURS AND INTUMESCENCE. By Dr. BRIGHT. Reprinted from the "Guy's Hospital Reports," with a Preface by Dr. BARLOW. Numerous Woodcuts.

A CLINICAL ACCOUNT OF DISEASES OF THE LIVER. By Prof. FRERICHS. 2 vols. Translated by Dr. MURCHISON. Coloured Lithographs, and numerous Woodcuts.

CZERMAK ON THE PRACTICAL USES OF THE LARYNGOSCOPE. Translated by Dr. G. D. GIBB. Numerous Woodcuts.

A HAND-BOOK OF PHYSICAL DIAGNOSIS COMPRISING THE THROAT, THORAX, AND ABDOMEN. By Dr. PAUL GUTTMANN, of Berlin. Translated by Dr. NAPIER, of Glasgow.

"We are persuaded that if the practitioner will carefully study this work, and conscientiously carry out its suggestions, he will find an incalculable advance in the realistic appreciation of diseases by means of their physical phenomena. The work is not properly a 'students' book. It presumes a certain familiarity with the diseases of the organs with which it deals, and the endeavour is made to connect the physical phenomena with the pathological conditions present in these diseases. . . . . . It was a wise decision of the New Sydenham Society to place a translation of it in the hands of their subscribers."—*Glasgow Medical Journal*, March, 1880.

"The New Sydenham Society has done well to put within the reach of their subscribers a work which not only has attained to a third edition in its own language, but has also been translated into Italian, Russian, Spanish, French, and Polish. . . . . . As a systematic and scientific treatise it well repays perusal. The book concludes with a good account of laryngoscopy, and of the physical signs of the principal diseases of the larynx. The acoustics of percussion and auscultation are elaborated with great care, and the precise explanation of the causes of many familiar physical signs will be very acceptable to teachers of clinical medicine, who have hitherto felt the want of an adequate scientific exposition of the principles of physical diagnosis."—*Dublin Journal of Medical Science*, November, 1880.

AN ATLAS OF ILLUSTRATIONS OF PATHOLOGY, COMPILED (CHIEFLY FROM ORIGINAL SOURCES) FOR THE SOCIETY.

The Committee in charge of this work consists of Dr. GEE, Dr. GREEN, Dr. MOXON, Dr. SUTTON, Mr. HOLMES, and Mr. HUTCHINSON.

EIGHT FASCICULI have been published, and it is proposed to issue one every year.

"Of the many valuable works published by this great Society, none are more acceptable to us than the 'Atlas of Pathology.'.... Such a vast and desirable undertaking as the publishing of this work is worthy of the Society named after the greatest English physician."—*Medical Press and Circular*, August 14, 1889.

The following subjects have been illustrated :—

FIRST FASCICULUS.

Scrofula; Syphilis; and Lymph-Adenoma.—Plate I.

Fig. 1. Scrofulous Disease of the Kidney and Ureter. Fig. 2. Scrofulous Disease of the Kidney. Fig. 3. Scrofulous Disease of the Kidney. Fig. 4. A Mass of Syphilitic Deposit in the Cortical Substance of the Kidney. Fig. 5. Lymph-Adenoma of Kidney.

Nephritis after Diphtheria; Scarlet Fever; and Burns.—Plate II.

Fig. 1. Nephritis after Diphtheria.—Section of Kidney. Fig. 2. Subacute Nephritis after Scarlet Fever.—Outer surface of kidney. Fig. 3. Subacute Nephritis after Scarlet Fever. Fig. 4. Acute Nephritis after Scarlet Fever. Fig. 5. Subacute Nephritis after Scarlet Fever. Fig. 6. Acute Nephritis after a Burn.—Outer surface of the kidney of a child who died after a very extensive burn. Fig. 7. Acute Nephritis after a Burn.—Section of the same kidney.

The Granular Kidney in different stages.—Plate III.

Fig. 1. Extremely Granular Kidney. Fig. 2. Extremely Granular Kidney.—Section of the same kidney. Fig. 3. Less Granular (contracted) Kidney.—Outer surface of the right kidney taken from the same subject as the left kidney shown in Figs. 1 and 2. Fig. 4. Granular Kidney of Bright. Fig. 5. Contracted Granular Kidney, in section. Fig. 6. Contracted Granular Kidney; exterior. Fig. 7. Large Granular Kidney. Fig. 8. Large Granular Kidney with cysts.

Embolism; Infarction Processes from Pyæmia; Jaundice and Purpura; Scrofula.—Plate IV.

Fig. 1. Embolic Changes in Pyæmia. Fig. 2. Embolic Changes in Pyæmia. Fig. 3. Pyæmic Deposits in Kidney. Fig. 4. Pyæmic Deposits in the Kidney. Fig. 5. Results of Jaundice and Purpura. Fig. 6. A variety of the Scrofulous Kidney.—The substance of the kidney is wholly destroyed and replaced by cavities containing a white mortar-like substance.

SECOND FASCICULUS.

Diseases of the Kidney.—Plate V.

Fig. 1. Amyloid Disease of Kidney in advanced stage. Fig. 2. A section of the same Kidney. Fig. 3. The pale flabby Kidney. Fig. 4. The same organ seen in section. Fig. 5. Medullary Cancer of the Kidney.

Various Diseased Conditions of the Spleen.—Plate VI.

Fig. 1. Hodgkin's Disease of Spleen (Lympho-sarcoma). Fig. 2. Acute Splenic enlargement in Diphtheria. Fig. 3. Suppurating infarction of Spleen from a case of Ulcerative Endocarditis. Fig. 4. Embolic changes in Pyæmia. Fig. 5. Rupture of the Spleen.

Diseases of the Supra Renal Capsules and Spleen.—Plate VII.

Fig. 1. Cancer of the Supra Renal Capsule. Figs. 2, 3, 4. Adenoma of the Supra Renal Capsule. Fig. 6. Addison's Disease of the Supra Renal Capsule (in section). Fig. 5. Addison's Disease of the Supra Renal Capsule.—"Fibro-calcareous or strumous disease." Fig. 7. Tubercle of the Spleen (external surface). Fig. 8. Tubercle of the Spleen (in section). Fig. 9. Lardaceous Spleen.

Microscopic Pathology of Kidneys.—Plate VIII.

Fig. 1. Lardaceous Degeneration of the Kidney.—Section of cortex. Fig. 2. Lardaceous Degeneration.—g. A glomerulus from the same kidney, as in Fig. 1, which has undergone lardaceous degeneration and is becoming fatty. Fig. 3. Part of the same seen with a higher power, showing contents of one of the tubules. Fig. 4. Lardaceous Degeneration in earlier stage combined with interstitial fibrous change. Figs. 5 & 6. Lardaceous Degeneration (after Cornil). Fig. 5. Section showing the hyaline membranous wall of the tubules *a a* much swollen, stained violet-red, showing waxy degeneration. Fig. 6. Transverse section of one of the pyramids, near summit of cone. Fig. 7. Granular Contracted Kidney. Fig. 8. From the same.—A thickened arteriole surrounded by fibroid growth. Fig. 9. Partial Fibrous Degeneration of Malpighian body in slight chronic intertubular nephritis. Fig. 10. From the same kidney; showing early changes around Malpighian body. Fig. 11. Multiplication of Nuclei on glomerulus with adhesion of capillary tuft to wall of capsule. Fig. 12. Subacute Interstitial Nephritis with large white kidney. Fig. 13. Scarlatinal Nephritis.—Intertubular exudation in a case fatal on 7th day of fever. Fig. 14. Subacute Interstitial Nephritis. Fig. 15. Acute Catarrhal Nephritis, showing swelling and granular degeneration of epithelium. (100 diam.) Fig. 16. Part of the same seen with a higher power. Fig. 17. Section of cortex from a case of parenchymatous (catarrhal) nephritis at a later stage (so-called "fatty" kidney). Fig. 18. From nearly transverse section near base of pyramid in similar case. Fig. 19. Casts in tubes in interstitial nephritis (post scarlatinal). Fig. 20. Colloid cast, *b*, in tubule; *a*, unaltered epithelium.

Microscopic Pathology of the Kidney.—Plate IX.

Fig. 1. Scarlatinal Nephritis. Fig. 2. Shows two of the glomeruli from same section as Fig. 1. Fig. 3. Section from the same.—Part of the wall of a Malpighian body from which the capillary tuft has fallen out. Fig. 4. Scarlatinal Nephritis.—(From a case fatal about 12 weeks from attack of fever). Fig. 5. Scarlatinal Nephritis. —(From a case fatal 15 months after attack of scarlet fever). Fig. 6. From same kidney as Fig. 5, but in a deeper part of cortex, close to medulla. Similar growth of interstitial connective tissue. Fig. 7. Subacute Interstitial Nephritis, probably Scarlatinal, under low power; showing diffuse infiltration and cluster of dilated tubules. Fig. 8. Chronic Parenchymatous Nephritis (large white kidney) with little or no interstitial change.—Section of cortex, showing changes in epithelium of convoluted tubules. Fig. 9. Kidney in leucocythæmia—to show localisation of changes around glomeruli and vessels. Fig. 10. Swelling of inner coat of small artery in granular contracted kidney. Fig. 11. Tuberculous Pyelonephritis. Fig. 12. Fatty Degeneration from Alcoholic Poisoning (after Lancereaux). Fig. 13. Fatty Degeneration in Cancer. Fig. 14. Individual epithelial cells from the preceding section; in various stages of fatty degeneration. Fig. 15. Cystic Degeneration of Kidney (after Lancereaux). Fig. 16. From a cyst in kidney near base of pyramid. Fig. 17. Colloid Degeneration of Kidney. Figs. 18, 19, 20, and 21, illustrate the hyaline changes found in the splenic arteries in certain febrile conditions. Fig. 18. From a section through the spleen of a case of early scarlatina, showing hyaline degeneration of the coat of an artery, transversely cut. Fig. 19.

Artery in longitudinal section. Fig. 20. Malpighian corpuscle from the spleen of a case of early scarlatina. Fig. 21. Part of the central and intermediate zone of the same Malpighian corpuscle as in Fig. 20, only more highly magnified (180 diam.) Fig. 22. Hodgkin's Disease. — Section of a spleen to show the overgrowth of the lymphatic sheath in Hodgkin's disease. (1 inch.) Fig. 23. Adenoma of the Supra Renal Capsule, showing the columns stuffed with fatty granules.

Microscopic Pathology of Spleen and Supra Renals.—Plate X.

Fig. 1. Capsulitis of the Spleen.—Vertical section of fibrous nodule in the capsule of the spleen, showing that the thickening of the capsule takes place by cellular growth in its deeper layers. Fig. 2. Fibrosis of the Spleen.—From the enlarged spleen of a ricketty child. Fig. 3. Fibrosis of the Spleen.—Showing a more advanced or fibrous condition spreading round some dilated veins. Fig. 4. Muscular Hypertrophy.—Over-growth of muscular trabeculæ in the spleen. Fig. 5. Muscular Hypertrophy. — Extreme stage of fibro-muscular growth in the spleen. Fig. 6. The Leucocythæmic Spleen.—Section of the edge of a Malpighian corpuscle, showing the compressed fibrous tissue between it and the splenic pulp. Fig. 7. The Leucocythæmic Spleen. — The pulp and stroma are normal. Fig. 8. Hodgkin's Disease.—The texture of a lymphoid nodule in the spleen of Hodgkin's disease. Fig. 9. Tubercular Spleen. (37 diam.) Fig. 10. Tubercular Spleen. Fig. 11. Induration and Atrophy. — A section of the spleen from a case of heart disease. Fig. 12. Lardaceous Spleen.—The sago spleen, showing the Malpighian corpuscles and small arteries mapped out by structureless hyaline lardaceous matter. Fig. 13. Lardaceous Spleen.—Transverse section of a Malpighian corpuscle, or small artery, with its surrounding lymphoid sheath. Fig. 14. Addison's Disease.—Vertical section of a supra renal capsule from the exterior inwards, to show the early changes in Morbus Addisonii. (250 diam.) Fig. 15. Addison's Disease.—Section of a supra renal capsule, to show the late, or fibro-calcareous, stage of Morbus Addisonii.

With Essay on the Pathology of the Kidney, by Dr. Greenfield. Essay on the Pathology of the Spleen and Supra Renals, by Dr. Goodhart.

"We look on this Pathological Atlas, in all its three fasciculi, as one of the best things that the Society has as yet done. The illustrations are nearly life size; the colouring is beautiful and true to nature; and we have not seen in this or any other country any work of this kind that satisfied us so much. Taken alone, it would be well worth the annual guinea; and will, when finished, constitute a treatise which every practising physician should possess."
—*Medical Press and Circular*, June 22nd, 1881.

THIRD FASCICULUS.

Diseases of the Liver.—Plate XI.

Lymphadenoma of Liver.

Diseases of the Liver.—Plate XII.

Fig. 1. Dilatation of the Bile Ducts in the Liver from pressure of a gall stone in cystic duct.

Fig. 2. Cancer of the Liver, with dilatation of the ducts and staining of the hepatic tissue.

Diseases of the Liver.—Plate XIII.

Syphilitic Cirrhosis of the Liver.

Diseases of the Liver.—Plate XIV.

Fig. 1. Red Atrophy, with acute Yellow Atrophy of the Liver.
Fig. 2. Microscopical appearances of the yellow swollen parts of the Liver (Acute Yellow Atrophy).
Fig. 3. Microscopical appearances of Red Atrophy of the Liver.

Diseases of the Liver.—Plate XV.

Fig. 1. Lardaceous Liver.
Fig. 2. Lardaceous Liver, showing the iodine reaction.

Diseases of the Liver.—Plate XVI.

Fig. 1. Cancer of the Liver.
Fig. 2. Nutmeg Liver, Chronic Congestion, and Atrophy of the Liver from mitral disease.

FOURTH FASCICULUS.

Diseases of the Liver, including one Figure of Spleen.—Plates XVII. to XXII.

Diseases of the Liver and Spleen.—Plate XVII.

Fig. 1. Cirrhosis of the Liver resembling the Nutmeg Liver.
Fig. 2. Brown Atrophy of the Liver.
Fig. 3. Cirrhosis of the Liver.
Fig. 4. Lymphadenoma of the Spleen (Hodgkin's Disease).

Diseases of the Liver.—Plate XVIII.

Fig. 1. Fatty Liver from Poisoning by Phosphorus.
Fig. 2. Cirrhosis of the Liver
Fig. 3. Tubercular Liver.
Fig. 4. Cirrhosis of the Liver.

Diseases of the Liver.—Plate XIX.

Cystic Disease of the Liver.

Diseases of the Liver.—Plate XX.

Fig. 1. Lardaceous Disease of the Liver. Fig. 2. Fatty Liver. Fig. 3. Early Cirrhosis. Figs. 4 & 5. Cirrhosis of the Liver (after Hamilton). Fig. 6. Cirrhosis of the Liver. Fig. 7. A Vegetation from the surface of the Liver. Fig. 8. Spindle-cell Sarcoma of the Liver. Fig. 9. Disseminated Growths of Fibrous Nature in the Liver. Fig. 10. Lardaceous Disease of the Liver. Fig. 11. Cavernous Tumour in the Liver. Fig. 12. Acute Yellow Atrophy of the Liver. Fig. 13. Cavernous Tumour in the Liver. Fig. 14. Early Cirrhosis. Fig. 15. Columnar Epithelioma of the Liver.

Diseases of the Liver.—Plate XXI.

Fig. 1. Cirrhosis of the Liver. Fig. 2. Cirrhosis of the Liver. Fig. 3. Monolobular Cirrhosis. Fig. 4. The Nutmeg Liver (Romose Atrophy of Moxon). Fig. 5. Tubercular Liver. Fig. 6. The Nutmeg Liver. Fig. 7. Miliary Gummata. Fig. 8. Idiopathic Anæmia. Figs. 9 & 10. Cancer of the Bile Ducts. Fig. 11. Cancer spreading from the Biliary Ducts. Fig. 12. Early Gummatous Infiltration of the Liver. Fig. 13. "Common" Cirrhosis. Fig. 14. Tubercular Liver. Fig. 15. Idiopathic Anæmia.

Diseases of the Liver.—Plate XXII.

Fig. 1. "Pericellular" Cirrhosis. Fig. 2. Cirrhosis of the Liver. Fig. 3. Nutmeg Liver. Fig. 4. Cystic Liver. Fig. 5. Cystic Liver. Fig. 6. Early Cancer of the Liver. Fig. 7. Extreme Tubercular Disease of the Liver. Fig. 8. Brown Atrophy of the Liver. Fig. 9. Extreme Tubercular Disease. Fig. 10. Myxœdematous Liver. Figs. 11, 12 & 13. "Contracting Scirrhus of the Liver simulating Cirrhosis." Figs. 14, 15 & 16. Varieties of Cell Vacuolation and Proliferation. Fig. 17. Primary Adenoma of the Liver. Fig. 18. Leukæmic Liver. Fig. 19. Primary Adenoma of the Liver.

FIFTH FASCICULUS.

Diseases of the Liver (chiefly of the Gall-Bladder and Larger Bile Ducts).—Plate XXIII.

Syphilitic and Lardaceous Disease of the Liver.

Diseases of the Liver.—Plate XXIV.

Fig. 1. Abscesses in the Liver.
Fig. 2. Papilloma of the Gall-Bladder.

Diseases of the Liver.—Plate XXV.

Cancer of Gall-Bladder and Liver.
Gall-stones, with Obstruction and Dilatation of the Cystic Duct.

Diseases of the Liver.—Plate XXVI.

Cancer of the Stomach extending to the Cystic Duct.

"We have nothing but praise to bestow on these plates, which are wonderfully good, and well worth the whole guinea subscription."—*Medical Press*, August 29, 1883.

SIXTH FASCICULUS.

Hydatid Cysts of the Liver.—Plate XXVII.
Urinary Calculi.—Plates XXVIII. to XXXI.
Comprising 46 Figures.

"Of the many valuable works published by this great Society, none are more acceptable to us than this Atlas of Pathology, of which we have received the sixth fasciculus. Such a vast and desirable undertaking as the publishing of this work is worthy of the Society named after the greatest English physician. . . . . We think that no medical man will be consulting his best interests if he hesitates to become a member of the Society. He certainly will have no more useful books than those bearing the medallion of the immortal Sydenham."—*Medical Press and Circular.*

SEVENTH FASCICULUS.

Urinary Calculi and Gall Stones.—Plate XXXII.
Enlargement of the Prostate Gland.—Plate XXXIII.
Enlargement of Prostate, Urinary Calculi.—Plate XXXIV.
Osteitis Deformans (Paget's Disease).—Plate XXXV.
Comprising 80 Figures.

EIGHTH FASCICULUS.

Diseases of Brain and Spinal Cord.—Plates XXXVI.-XLI.

Plate XXXVI.

Fig. 1. Hydatid in the Posterior Corner of the Right Lateral Ventricle.
Fig. 2. Abscess on the Under Surface of the Right Cerebellar Hemisphere, close to the Petrous Portion of the Temporal Bone.

Plate XXXVII.

Fig. 1. Hæmorrhage into the Right Hemisphere and Median Lobe of the Cerebellum.
Fig. 2. Tubercles of various sizes situated on the Upper Surface of the Cerebellar Hemispheres.
Fig. 3. A Tuberculous Tumour situated between the left side of the Pons Varolii, the Medulla Oblongata, and the adjacent surface of the Cerebellar Hemisphere.

Plate XXXVIII.

Fig. 1. A severely crushed Spinal Cord.
Fig. 2. The Cervical Spinal Cord of a Man who had died under almost precisely similar conditions to those specified in the preceding case.
Fig. 3. Hæmorrhage external to the Vertebral Theca.

Plate XXXIX.

Figs. 1, 2, & 3. A Tuberculous Tumour on the Spinal Dura Mater.

Plate XL.

Fig. 1. Cartilaginous Deposits on the Spinal Arachnoid.
Fig. 2. Myelitis after Concussion of the Spine.

Plate XLI.

Fig. 1. Tubercle in Pia Mater of Cord.
Fig. 2. A Fibrous Tumour lodged in the Cauda Equina.

ON THE DISEASES OF OLD AGE. By Prof. CHARCOT. Translated by Mr. WILLIAM TUKE.

"The New Sydenham Society has been well advised in presenting to its readers one of the most important neurological works which has appeared of late years. . . . . . Charcot's volume is a book to read and re-read for all of us."—*British Medical Journal*, June 23, 1883.

THE DIAGNOSIS AND TREATMENT OF DISEASES OF THE CHEST. By Dr. STOKES. A Reprint Edited by Dr. HUDSON, of Dublin.

"His fame as one of the foremost physicians of his age may, we think, rest securely upon his two main works—on 'Diseases of the Chest,' and on 'Diseases of the Heart and Aorta.' Each of these treatises is to be reckoned a 'historical landmark in medicine,' and it was from this point of view, as it would seem, that Dr. Hudson undertook the editing of the volume under notice. . . . . . . Prefixed to this edition is a graceful and sympathetic memoir of Dr. Stokes, by his attached friend, Dr. Ackland. It would be an impertinence to enter into any detailed notice of a book which is universally esteemed one of the Medical Classics, and it is sufficient to point out the principal changes which have been made. . . . . . . The reprint of this work tells its own tale."—*Dublin Journ. of Med. Science.* Oct., 1883.

THE COLLECTED WORKS OF DR. WARBURTON BEGBIE. Edited by Dr. DYCE DUCKWORTH. With a Memoir and Portrait.

"The Council of the Sydenham Society, and Dr. Duckworth, in particular, have done a good work in collecting these writings together into a volume, and the profession in Scotland and in many places beyond will feel grateful to them for this memorial of one who lives in the affectionate remembrance and admiration of his professional brethren and associates, and who, indeed, had earned in the public esteem the title of the 'beloved physician.'"—*Edinburgh Medical Journal*, June, 1883.

SELECTIONS FROM THE CLINICAL WORKS OF DR. DUCHENNE (of Boulogne). Translated and Edited by Dr. VIVIAN POORE.

"We have, however, before us a volume produced under the auspices of the New Sydenham Society, the addition of which to an English standard library cannot but be looked upon as a national compliment to our illustrious confrere. ...... The arrangement of the work under notice has been carried out with judgment and in a concise yet comprehensive form. All the characteristic and important conclusions are presented to the English reader."—*Medical Times and Gazette*, Jan. 5, 1884.

"The work of condensation and selection appears to have been admirably performed; nearly all the material is drawn from Duchenne's great work, which bears the somewhat misleading title of 'L'Electrisation Localisée.'"—*British Medical Journal*, Jan. 12, 1884.

"The New Sydenham Society may well be congratulated on the appearance of such a valuable work in the English language. Duchenne's world-wide reputation on the subject of Nervous Diseases rendered it necessary that his life-long labours should be extensively known among English students and practitioners. The work before us has been ably translated, edited, and condensed by Dr. G. V. Poore."—*Medical Press and Circular*, April 16, 1884.

CLINICAL LECTURES ON THE PRACTICE OF MEDICINE. Vols. I. and II. By the late ROBERT J. GRAVES, M.D. Reprinted from the Second Edition, Edited by Dr. NELIGAN.

"The reprint of 'Graves' Practice of Medicine' once more places within the reach of every member of the profession a classical work by one of the greatest clinical teachers."—*British Medical Journal*, February 20, 1886.

"'Graves' Clinical Lectures' need no eulogium now, and, although more than forty years have passed since the first edition of them was published, they still maintain a worthy place among medical classics, and will amply repay a careful perusal."—*British Medical Journal*, June, 1886.

ALBUMINURIA IN HEALTH AND DISEASE. By Dr. H. SENATOR. Translated by Dr. T. P. SMITH.

"Dr. Senator on Albuminuria may be taken as the latest utterances of science upon this obscure and difficult subject, to the elucidation of which the author has contributed so much. ...... The translation faithfully rendered by Dr. T. P. Smith, and the monograph is one which requires to be carefully studied, for it is fully of closely-reasoned argument."—*Lancet*, June 20, 1885.

SOME CONSIDERATIONS ON THE NATURE AND PATHOLOGY OF TYPHUS AND TYPHOID FEVER. By the late Dr. P. STEWART, Edited by Dr. W. CAYLEY.

"Dr. Cayley has edited the late Dr. H. P. Stewart's paper on the Identity or Non-identity of Typhus or Typhoid Fever. It is satisfactory to find that a scientific production of such undoubted merit has been rescued from oblivion. —*Lancet*, June 20, 1885.

Surgery.

ESMARCH ON THE USES OF COLD IN SURGICAL PRACTICE. Translated by Dr. Montgomery. Woodcuts.

"Esmarch's treatise is of high practical interest."—*British Medical Journal.*

BILLROTH'S LECTURES ON SURGICAL PATHOLOGY AND THERAPEUTICS. A Hand-book for Students and Practitioners. 2 vols.

INVESTIGATION INTO THE ETIOLOGY OF THE TRAUMATIC INFECTIVE DISEASES. By R. Koch. Translated, with Lithographic Plates, by Mr. Watson Cheyne.

ON THE PROCESS OF REPAIR AFTER RESECTION AND EXTIRPATION OF BONES. By Dr. A. Wagner, of Berlin. Translated by Mr. T. Holmes.

CLINICAL LECTURES. Selected from Professor Volkmann's Series. 2 vols. (See "Medicine.")

THE WORKS OF ABRAHAM COLLES. Chiefly his Treatise on the Venereal Disease and on the Use of Mercury. Edited, with Portrait, by Dr. McDonnell, of Dublin.

Gynæcology and Midwifery.

A TEXT-BOOK OF MIDWIFERY. By Otto Spiegelberg. Translated from the Second German Edition by Dr. J. B. Hurry. 2 vols.

"The reputation attained by the late author of this work suffices to explain its translation and publication by the New Sydenham Society. Written in a lucid and easy style, it leads the reader on indefinitely without ever causing weariness. . . . . The book is well and profusely illustrated, and should find a place in the general practitioner's library. Very few English text-books describe treatment as this does. The translation is everything that can be desired."—*London Medical Record.*

"The issue of the second volume of Spiegelberg's Midwifery, translated by Dr. Hurry, under the auspices of the New Sydenham Society, now places this valuable work in a complete form within the reach of the English reader who is not familiar with German. . . . . It would be difficult to speak too highly of the book as a general text-book of Midwifery; it is neither too long nor too difficult for the student, while the practitioner wishing to read up the authorities on some special point will find ample information of a practical kind. Dr. Hurry is to be congratulated on the way he has done his work. The book is commendably free from idiomatic indications of its foreign origin."
—*British Medical Journal*, Oct. 5, 1889.

"The New Sydenham Society has done much good work and conferred many benefits upon the English student and practitioner, but it has not done better than by giving Spiegelberg's great book on Midwifery to the English reader."—*Lancet*, Jan. 7, 1888.

HISTORY AND ETIOLOGY OF SPONDYLOLISTHESIS. By Dr. FRANZ LUDWIG NEUGEBAUER, of Warsaw. Translated by Dr. FANCOURT BARNES.

ON THE MORE IMPORTANT DISEASES OF WOMEN AND CHILDREN, with other Papers, by Dr. GOOCH. Reprinted; with a Prefatory Essay by Dr. ROBERT FERGUSON. With woodcuts.

CLINICAL MEMOIRS ON DISEASES OF WOMEN. By Drs. BERNUTZ and GOUPIL. 2 vols. Translated and abridged, Dr. MEADOWS.

"The careful study of these valuable memoirs is imperative on all who are interested in gynæcology."—*Lancet*, October, 1866.

MOVEABLE KIDNEY IN WOMEN. By Dr. LEOPOLD LANDAU. Translated and Edited, with notes, by FRANCIS HENRY CHAMPNEYS, M.A.

"The essay of Dr. Leopold Landau on Moveable Kidney in Women fills a gap in medical literature. It deals with the subject in the most thorough manner; but space forbids our attempting any analysis of the work, for the translation of which we are indebted to Dr. Champneys."—*Lancet*, June 20, 1885.

TREATISE ON THE THEORY AND PRACTICE OF MIDWIFERY. 3 vols. Edited and Annotated by Dr. McCLINTOCK, of Dublin. With Portrait of SMELLIE.

"This book begins with a fine engraving of the author, and had the New Sydenham Society done for Smellie's memory no more than the publication of this valuable print, it would have a strong claim on the gratitude of the profession. McClintock's life of Smellie is a very interesting contribution to medical literature. His works show that he was a very great man and midwife, but his biography was needed to show his peculiarities. . . . . . Let the reader carefully peruse Dr. McClintock's annotations, and he will see how Smellie's Editor recognises Smellie's keenness of eye in discerning how to make progress."
—*Edin. Med. Journal*, March, 1877.

Diseases of the Eye and Ear.

ON THE ANOMALIES OF ACCOMMODATION AND REFRACTION OF THE EYE, with a PRELIMINARY ESSAY ON PHYSIOLOGICAL DIOPTRICS. By F. C. DONDERS, M.D., Professor of Physiology and Ophthalmology in the University of Utrecht. Written expressly for the Society. Translated from the Author's Manuscript by W. D. MOORE, M.D.

"This splendid monograph, from the hand of the accomplished professor of physiology and ophthalmology, of Utrecht, will be hailed as a boon by all lovers of ophthalmic science."—*Lancet.*

THREE MEMOIRS ON GLAUCOMA AND ON IRIDECTOMY AS A MEANS OF TREATMENT. By Professor VON GRÆFE. Translated by Mr. T. WINDSOR, of Manchester.

ON THE MECHANISM OF THE BONES OF THE EAR AND THE MEMBRANA TYMPANI. (Pamphlet.) By Professor HELMHOLTZ. Translated by Mr. HINTON.

THE AURAL SURGERY OF THE PRESENT DAY. By W. KRAMER, M.D., of Berlin. Translated by HENRY POWER, Esq., F.R.C.S., M.B. With two Tables and nine Woodcuts.

VON TROELTSCH'S TREATISE ON DISEASES OF THE EAR. Translated, with Notes, by Mr. HINTON.

Forensic Medicine.

A HANDBOOK OF THE PRACTICE OF FORENSIC MEDICINE, BASED UPON PERSONAL EXPERIENCE. By J. L. CASPER, M.D., late Professor of Medical Jurisprudence in the University of Berlin. Translated by G. W. BALFOUR, M.D. 4 vols.

"Casper's great work, based as it is upon a minute and laborious observation of facts, must prove the most trustworthy guide in the interpretation of the ofttimes difficult questions which the medical jurist is called upon to solve."—*Lancet.*

"This work must be regarded as a valuable and judicious addition to the publications of the Society from which it emanates. The advantages to be derived by the reader from its perusal cannot be over-estimated or too eagerly sought for."—*Madras Quarterly Journal of Medical Science.*

Diseases of the Nervous System.

SCHRŒDER VAN DER KOLK ON A CASE OF ATROPHY OF THE LEFT HEMISPHERE OF THE BRAIN. Translated by Dr. W. MOORE, of Dublin. Four Lithographs.

ON THROMBOSIS OF THE CEREBRAL SINUSES By Professor VON DUSCH. Translated by Dr. WHITLEY.

LECTURES ON DISEASES OF THE NERVOUS SYSTEM. By Professor CHARCOT. (First, Second, and Third Series.) Translated by Dr. SIGERSON, of Dublin. With woodcuts.

"These lectures of M. Charcot are too well known in the original to call for any special criticism here. They have, indeed, obtained an European reputation, and it has long been felt that it would be a great gain to our literature to have them rendered into English. . . . . . We strongly advise all those of our readers who may not yet have made themselves acquainted with these lectures to lose no time in doing so. . . . . . The translator, Dr. Sigerson, a former pupil of the author, has succeeded admirably in his rendering of the elegant literary style of M. Charcot. . . . . . It is, without doubt, one of the most valuable books that has been issued by this Society since their translation of Trousseau."—*Lancet*, August, 1877.

"This volume will be highly prized by the members of the N. S. S. M. Charcot's name ranks among the very foremost of those who have advanced the knowledge of nerve-pathology. The work he has done is marked by great accuracy and close observation, and by great acumen in interpreting facts and drawing inferences."—*Brit. and For. Med. Chir. Rev.*, July, 1877.

A MANUAL OF MENTAL PATHOLOGY AND THERAPEUTICS. By Professor GRIESINGER. Translated by Dr. LOCKHART ROBERTSON and Dr. JAMES RUTHERFORD.

ON EPILEPSY. By Professor SCHRŒDER VAN DER KOLK.

CHARCOT'S TREATISE ON THE LOCALISATION OF CEREBRAL AND SPINAL DISEASE. Translated by Dr. HADDEN.

"It will give to its reader a clear understanding of what is known of the subject it professes to treat of."—*Edinburgh Medical Journal*, Dec., 1883.

"Dr. Hadden is to be congratulated upon having produced a translation of these valuable Lectures, which, whilst faithful to the text, is not marred by being too literal."—*Medical Times and Gazette*, October 27, 1883.

"We do not attempt to give a detailed notice of M. Charcot's views and descriptions. Our object in this short notice will be secured if we induce our readers to possess themselves of the work itself."—*Journal of Medical Science*, July, 1884.

"Still we must admire the skill with which M. Charcot utilises anatomical facts as aids to clinical diagnosis, and we are persuaded that these Lectures are well worthy of attentive study by all who care to dig below the surface in the investigation of nervous diseases."—*Dublin Journ. of Med. Seience*, July, 1884.

Anatomy, Physiology, and General Pathology.

A MANUAL OF HUMAN AND COMPARATIVE HISTOLOGY. By S. STRICKER. 3 vols. Translated by Mr. POWER.

"This work, edited by Stricker, and having as its contributors nearly all of the best names in Germany, is one well deserving of attention, and constitutes, we think, a very valuable addition to the stores of the New Sydenham Society." —*Medical Times and Gazette*, December 10, 1870.

EXPERIMENTAL RESEARCHES ON THE EFFECTS OF LOSS OF BLOOD IN PRODUCING CONVULSIONS, By Drs. KUSSMAUL and TENNER. Translated by Dr. BRONNER, of Bradford.

A MANUAL OF PATHOLOGICAL HISTOLOGY, intended to serve as an introduction to the study of Morbid Anatomy. By Professor RINDFLEISCH. (Bonn.) 2 vols. Translated by Dr. BAXTER.

"The members of the Society may be congratulated on the addition of such valuable treatises to their libraries. . . . . . The Society ought to flourish whilst it caters so well for its members. They have every reason to be content both with the quantity and quality of the matter supplied." — *Brit. and For. Chir. Rev.*, July, 1873.

AN ATLAS OF ILLUSTRATIONS OF PATHOLOGY. (See "Medicine," page 10.)

ON THE MINUTE STRUCTURE AND FUNCTIONS OF THE SPINAL CORD. By Professor SCHRŒDER VAN DER KOLK. Translated by Dr. W. D. MOORE. Numerous Lithographs.

ON THE MINUTE STRUCTURE AND FUNCTIONS OF THE MEDULLA OBLONGATA, AND ON EPILEPSY. By Professor SCHRŒDER VAN DER KOLK. Translated by Dr. W D. MOORE. Numerous Lithographs.

Retrospects, and Works of General Reference.

A YEAR-BOOK OF MEDICINE AND SURGERY, AND THEIR ALLIED SCIENCES, for 1859. Edited by Dr. HARLEY, Dr. HANDFIELD JONES, Mr. HULKE, Dr. GRAILY HEWITT, and Dr. ODLING.

"Our space will not admit of a further statement of the excellent character of the Year-Book and the other works issued by the New Sydenham Society, but we should strongly urge every member of the profession, who has the advancement of medical knowledge at heart, to lose no time in forwarding his name, should he not already have done so."—*London Medical Journal.*

YEAR-BOOK for 1860. Edited by Dr. HARLEY, Dr. HANDFIELD JONES, Mr. HULKE, Dr. GRAILY HEWITT, and Dr. SANDERSON.

" This is, as it professes to be, an improvement on its predecessor. On the whole the editors have done their laborious work well."—*British Medical Journal*, December 31, 1861.

YEAR-BOOK for 1861. Edited by Dr. HARLEY, DR. HANDFIELD JONES, Mr. HULKE, Dr. GRAILY HEWITT, and Dr. SANDERSON.

YEAR BOOK for 1862. Edited by Dr. MONTGOMERY, Dr. HANDFIELD JONES, Mr. WINDSOR, Dr. GRAILY HEWITT, and Dr. SANDERSON.

YEAR-BOOK for 1863. By the same Editors.

YEAR-BOOK for 1864. Edited by Mr. HINTON, Dr. HANDFIELD JONES, Mr. WINDSOR, Dr. M. BRIGHT, and Dr. HILTON FAGGE.

A BIENNIAL RETROSPECT OF MEDICINE, SURGERY, AND THEIR ALLIED SCIENCES, for the Years 1865 and 1866. Edited by Mr. POWER, Dr. ANSTIE, Mr. HOLMES, Dr. BARNES, Mr. WINDSOR, and Dr. HILTON FAGGE.

A BIENNIAL RETROSPECT OF MEDICINE, SURGERY, AND THEIR ALLIED SCIENCES, for the Years 1867 and 1868. Edited by Mr. H. POWER, Dr. ANSTIE, Mr. HOLMES, Mr. R. B. CARTER, Dr. BARNES, and Dr. THOMAS STEVENSON.

A BIENNIAL RETROSPECT for 1869 and 1870.

A BIENNIAL RETROSPECT for 1871 and 1872.

A BIENNIAL RETROSPECT for 1873 and 1874.

" Full justice is done to English observers, and the whole volume is creditable to its compilers and to the Society under whose auspices it is published."—*Lancet*, January, 1876.

THE MEDICAL DIGEST. Being a means of ready reference to the principal contributions to Medical Science during the last Thirty years. By Dr. RICHARD NEALE.

BIBLIOTHECA THERAPEUTICA; OR BIBLIOGRAPHY OF THERAPEUTICS. By E. J. WARING, M.D. 2 vols.

"The preparation of such a catalogue as the present one must have entailed enormous labour, such as few men are capable of, and such as rarely brings them the thanks they deserve. We are quite sure, however, that all those who are engaged in the study of *Materia Medica*, who are not satisfied with merely looking at a few recent papers, but are desirous of learning all that has been done regarding the particular drug which may be the object of their attention, will be exceedingly grateful to Dr. Waring for sparing them so much labour."—*Practitioner*, December 18th, 1879.

A LEXICON OF MEDICAL TERMS. Edited by Mr. POWER and Dr. SEDGWICK. Parts I. to XVII. This Lexicon is based upon the well-known work of Dr. MAYNE, the copyright of which was purchased by the Society. It is, however, entirely rewritten by the present Editors, and very much enlarged.

"The work is carefully and elaborately done, and comprehends every reference which the medical or scientific inquirer could possibly require."— *Medical Press and Circular*, June 22nd, 1881.

"When complete, the work will be a most valuable addition to the library, not only of medical men, but of those scientists who are interested in Medicine and the allied sciences."—*R. Neale, M.D., in London Med. Recorder*, Feb. 1888.

"When finished, the Lexicon will be a credit to British Medicine, and worthy of the great Physician whose name the Society bears."—*Dublin Medical Journal*, Aug., 1890.

Diseases of the Skin and Syphilis.

ON SYPHILIS IN INFANTS. By PAUL DIDAY. Translated by Dr. WHITLEY.

ON DISEASES OF THE SKIN, INCLUDING THE EXANTHEMATA. By Professor HEBRA. 5 vols. Translated and Edited by Dr. HILTON FAGGE, Dr. PYE-SMITH, and Mr. WAREN TAY.

"Of all the works produced by the New Sydenham Society this is one of the most valuable and most welcome. . . . . . It is to be remarked that this book is not a mere translation of the German work; it is a new and revised edition, undertaken by the author for his English brethren."—*Medical Times and Gazette*, April 27, 1867.

"The New Sydenham Society has done good service to the medical profession by undertaking the translation and publication of Professor Hebra's excellent work. In several respects the English edition is greatly superior to the original. In closing its pages we have but one regret, namely, that the New Sydenham Society does not embody the whole medical confraternity, so that every member of our noble profession might have on his bookshelves a copy of this most valuable book."—*Journal of Cutaneous Medicine*, April, 1877.

LANCEREAUX'S TREATISE ON SYPHILIS. 2 vols. Translated by Dr. WHITLEY.

"The work is the most exhaustive book which has been published on the subject, and has been quoted by all the recent writers in this country, America, and the Continent. It is a perfect mine of information. The translation is well done, and the New Syd. Soc. may be congratulated on having added such an important treatise to its list of works."—*Lancet*, March, 1869.

The Society's Atlas of Diseases of the Skin.

In seventeen Fasciculi comprising the following subjects. Unless otherwise indicated, the Plates are original.

	PLATE
Favus. From Hebra.	I.
Tinea Tonsurans. From Hebra.	II.
Lupus Exulcerans. From Hebra.	III.
Psoriasis Diffusa. From Hebra.	IV.
Ichthyosis. From Hebra.	V.
Lupus Serpiginosus; Alopecia Areata. From Hebra.	VI.
Lupus Vulgaris et Serpiginosus (Cicatrising). From Hebra.	VII.
Herpes Zoster Frontalis (affecting the Frontal and Trochlear Branches of the Fifth Nerve).	VIII.
Molluscum Contagiosum, A, on a Child's Face; B, on the Breast of the Child's Mother; C, Anatomical Characters of the Tumours; D, Microscopic Characters.	IX.
Morbus Addisonii.	X.
Leucoderma.	XI.
Pemphigus.	XII.
Pityriasis Versicolor.	XIII.
Psoriasis Inveterata.	XIV.
Eczema Impetiginodes on Face of Adult.	XV.
Eczema on the Face, &c., of Infant; Eczema Rubrum on Leg of Adult.	XVI.
Psoriasis of Hands and Finger-nails; Syphilitic Psoriasis of Finger-nails; Congenito-Syphilitic Psoriasis of Finger- and Toe-nails; Onychia Maligna; Chronic General Onychitis..	XVII.
Molluscum Fibrosum seu Simplex.	XVIII.
Psoriasis-Lupus (Lupus non Exedens, in numerous Symmetrical Patches).	XIX.
Porrigo Contagiosa (e pediculis).	XX.
Erythema Nodosum..	XXI.
Morbus Pedicularis..	XXII.
Herpes Zoster (with scars of a former attack).	XXIII.
Erythema Circinatum.	XXIV.
Eczema (from Sugar).	XXV.
Acne Vulgaris..	XXVI.
Scabies (on Hand of Child). Scabies (with Œdema, &c.) Scabies Norvegica.	XXVII.
Porrigo Contagiosum after Vaccination. Circinate Eruptions in Congenital Syphilis.	XXVIII.
True Leprosy (Tubercular Form). True Leprosy (Anæsthetic Form).	XXIX.
Pityriasis Rubra.	XXX.

	PLATE
Papular Syphilitic Eruption, with Indurated Chancre on the Skin of the Abdomen.	XXXI.
Pruriginous Impetigo after Varicella.	XXXII.
Lichen of Infants.	XXXIII.
Kerion of Scalp after Ringworm.	XXXIV.
Eruption produced by Iodide of Potassium.	XXXV.
Tinea Circinata.	XXXVI.
Rupia-Psoriasis (from inherited Syphilis).	XXXVII.
Prurigo Adolescentium.	XXXVIII.
Purpura Thrombotica.	XXXIX.
Syphilitic Rupia, with Keloid on Scars	XL.
Framboesie (Endemic Verrugas).	XLI.
Lupus Erythematosus.	XLII.
Ulcerating Eruption from Bromide of Potassium.	XLIII
Morphæa, or Addison's Keloid.	XLIV.
Purpura Hæmorrhagica.	XLV.
Molluscum Contagiosum.	XLVI.
Pemphigus Foliaceus.	XLVII.
Inherited Syphilis.	XLVIII.
Syphilitic Tubercular Lupus.	XLIX.

"This Fasciculus supplies life size portraits of pityriasis rubra, papular syphilis, with indurated chancres, and pruriginous impetigo following varicella, which are extremely beautiful, and look life-like."—*Edin. Medical Journal*, May, 1872.

"They are better, to our mind, than any other plates in use amongst us; and there cannot be a question as to the Society's issue being as popular as it is useful."—*Lancet.*

"We have received the thirteenth fasciculus of this splendid collection of drawings, of which no further praise is needed than to say that they are executed with the same artistic skill and fidelity to nature which have characterised the whole series."—*Dublin Journal of Medical Science*, May, 1874.

A CATALOGUE OF THE PORTRAITS COMPRISED IN THE SOCIETY'S ATLAS OF SKIN DISEASES. Prepared, at the request of the Council, by Mr. HUTCHINSON. Parts 1 and 2.

"The descriptions, cases, and plates are well given. . . . . . There is one good feature in some of the cases described. Take that of Addison's Keloid, p. 160. In it we have notes, &c., of a rare skin disease, which has been accurately described by the observers under whose care the patient had been at various stages of the case. This is, therefore, a valuable contribution to medicine."—*Edinburgh Medical Journal*, February, 1877.

LIST OF PUBLISHED WORKS

Arranged according to the Year of Issue.

Vol. 1859. (*First Year.*)
1. Diday on Infantile Syphilis.
2. Gooch on Diseases of Women.
3. Memoirs on Diphtheria.
4. Van der Kolk on the Spinal Cord, &c.
5. Monographs (Kussmaul and Tenner, Graefe, Wagner, &c.)

1860. (*Second Year.*)
6. Dr. Bright on Abdominal Tumours.
7. Frerichs on Diseases of the Liver. Vol. I.
8. A Yearbook for 1859.
9. Atlas of Portraits of Skin Diseases. (1st Fasciculus.)

1861. (*Third Year.*)
10. A Yearbook for 1860.
11. Monographs (Czermak, Dusch, Radicke, &c.)
12.*Casper's Forensic Medicine. Vol. I.
14.*Atlas of Portraits of Skin Diseases. (2nd Fasciculus.)

1862. (*Fourth Year.*)
13. Frerichs on Diseases of the Liver. Vol. II.
15. A Yearbook for 1861.
16. Casper's Forensic Medicine. Vol. II.
17. Atlas of Portraits of Skin Diseases (3rd Fasciculus.)

1863. (*Fifth Year.*)
18. Kramer on Diseases of the Ear.
19. A Yearbook for 1862.
20. Neubauer and Vogel on the Urine.

VOL. 1864. (*Sixth Year.*)
21. CASPER's Forensic Medicine. Vol. III.
22.*DONDERS on Accommodation and Refraction of the Eye.
23. A YEARBOOK for 1863.
24. ATLAS of Portraits of Skin Diseases. (4th Fasciculus).

1865. (*Seventh Year.*)
25. A YEARBOOK for 1864.
26. CASPER's Forensic Medicine. Vol. IV.
27.*ATLAS of Portraits of Skin Diseases. (5th Fasciculus).

1866. (*Eighth Year.*)
28. BERNUTZ and GOUPIL on the Diseases of Women. Vol. I.
29. ATLAS of Portraits of Skin Diseases. (6th Fasciculus.)
30. HEBRA on Diseases of the Skin. Vol. I.
31. BERNUTZ and GOUPIL on Diseases of Women. Vol. II.

1867. (*Ninth Year.*)
32. BIENNIAL Retrospect of Medicine and Surgery.
33. GRIESINGER on Mental Pathology and Therapeutics.
34.*ATLAS of Portraits of Skin Diseases. (7th Fasciculus).
35. TROUSSEAU's Clinical Medicine. Vol. I.

1868. (*Tenth Year.*)
36. THE Collected Works of Dr. Addison.
37. HEBRA on Skin Diseases. Vol. II.
38. LANCEREAUX's Treatise on Syphilis. Vol. I.
39. ATLAS of Portraits of Skin Diseases. (8th Fasciculus).
40. CATALOGUE of Atlas of Skin Diseases. (First Part.)

1869. (*Eleventh Year.*)
41. LANCEREAUX's Treatise on Syphilis. Vol. II.
42. TROUSSEAU's Clinical Medicine. Vol. II.
43. BIENNIAL Retrospect of Medicine and Surgery.
44. ATLAS of Portraits of Skin Diseases. (9th Fasciculus.)

1870. (*Twelfth Year.*)
45. TROUSSEAU's Lectures on Clinical Medicine. Vol. III.
46. NIEMEYER's Lectures on Pulmonary Consumption.
47. STRICKER's Manual of Histology. Vol. I.
48. ATLAS of Portraits of Skin Diseases. (10th Fasciculus).

Vol.   1871.   (*Thirteenth Year.*)
49. WUNDERLICH's Medical Thermometry.
50. BIENNIAL Retrospect of Medicine and Surgery.
51. TROUSSEAU's Clinical Medicine. Vol. IV.
52. ATLAS of Portraits of Skin Diseases. (11th Fasciculus.)

1872.   (*Fourteenth Year.*)
53. STRICKER's Manual of Histology. Vol. II.
54. RINDFLEISCH's Pathological Histology. Vol. I.
55. TROUSSEAU's Clinical Medicine. Vol. V.
56. ATLAS of Portraits of Skin Diseases. (12th Fasciculus.)

1873.   (*Fifteenth Year.*)
57. STRICKER's Manual of Histology. Vol. III.
58. RINDFLEISCH's Pathological Histology. Vol. II.
59. BIENNIAL Retrospect of Medicine and Surgery.
60. ATLAS of Portraits of Skin Diseases. (13th Fasciculus.)

1874.   (*Sixteenth Year.*)
61. HEBRA on Skin Diseases. Vol. III.
62. Von TROELTSCH on Diseases of the Ear.
    HELMHOLTZ on Membrana Tympani, &c. (In one Vol.)
63. ATLAS of Portraits of Skin Diseases. (14th Fasciculus.)
64. HEBRA on Skin Diseases. Vol. IV.

1875.   (*Seventeenth Year.*)
65. BIENNIAL Retrospect of Medicine and Surgery.
66. CATALOGUE of Atlas of Skin Diseases. (Second Part.)
67. ATLAS of Portraits of Skin Diseases. (15th Fasciculus.)
68. CLINICAL Lectures by various German Professors. Vol. I.
69. LATHAM's Works. Vol. I.

1876.   (*Eighteenth Year.*)
70. SMELLIE's Midwifery, by McClintock. Vol. I.
71. CLINICAL Lectures by various German Professors. Vol. II.
72.*CHARCOT's Clinical Lectures on Diseases of the Nervous
      System. Vol. I.
73. BILLROTH's Lectures on Surgical Pathology. Vol. I.

1877.   (*Nineteenth Year.*)
74. SMELLIE's Midwifery, by McClintock. Vol. II.
75. THE Medical Digest, by Dr. Neale.
76. BILLROTH's Lectures on Surgical Pathology. Vol. II.
77. ATLAS of Illustrations of Pathology. (Fasciculus I.)

Vol.   1878. (*Twentieth Year.*)
78. BIBLIOTHECA Therapeutica, by Dr. Waring.   Vol. I.
79. SMELLIE's Midwifery, by McClintock.   Vol. III.
80. LATHAM's Works.   Vol. II.
81. LEXICON of Medical Terms.   (First Part.)   *Issued with Part II. only, as Vol. 83.*

1879. (*Twenty-first Year.*)
82. BIBLIOTHECA Therapeutica, by Dr. Waring.   Vol. II.
83. LEXICON of Medical Terms.   (Second Part.)   *Including re-issue of First Part.*
84. MANUAL of Physical Diagnosis, by Dr. Guttmann.
85. ATLAS of Illustrations of Pathology.   (Fasciculus II.)

1880. (*Twenty-second Year.*)
86. HEBRA on Diseases of the Skin.   Vol. V.
87. LEXICON of Medical Terms.   (Third Part.)
88. KOCH's Researches on Wound Infection.
89. LEXICON of Medical Terms.   (Fourth Part.)
90. CHARCOT's Clinical Lectures on Diseases of the Nervous System.   Vol. II.
91. ATLAS of Illustrations of Pathology.   (Fasciculus III.)

1881. (*Twenty-third Year.*)
92. SELECTIONS from the Works of Abraham Colles.
93. LEXICON of Medical Terms.   (Fifth Part.)
94. BILLROTH's Clinical Surgery.
95. CHARCOT on Diseases of Old Age.
96. LEXICON of Medical Terms.   (Sixth Part.)
97. ATLAS of Illustrations of Pathology.   (Fasciculus IV.)

1882. (*Twenty-fourth Year.*)
98. STOKES on Diseases of the Chest.
99. ATLAS of Portraits of Skin Diseases.   (16th Fasciculus.)
100. THE Collected Works of Dr. Warburton Begbie.
101. LEXICON of Medical Terms.   (Seventh Part.)
102. CHARCOT on Localisation of Cerebral and Spinal Disease.
103. LEXICON of Medical Terms.   (Eighth Part.)

1883. (*Twenty-fifth Year.*)
104. ATLAS of Illustrations of Pathology.   (Fasciculus V.)
105. SELECTIONS from the Works of Dr. Duchenne.
106. HIRSCH on Geographical and Historical Pathology. Vol. I.
107. LEXICON of Medical Terms.   (Ninth Part.)

1884. (*Twenty-sixth Year.*)

108. ATLAS of Portraits of Skin Diseases. (17th Fasciculus.)
109. GRAVES' Clinical Medicine. Vol. I. (Reprinted.)
110. SELECTED Monographs:— Senator on Albuminuria; Stewart on Typhus and Typhoid Fever; Landau on Moveable Kidney in Women.
111. LEXICON of Medical Terms. (Tenth Part.)

1885. (*Twenty-seventh Year.*)

112. HANDBOOK of Geographical and Historical Pathology. By Dr. Aug. Hirsch. Vol. II.
113. GRAVES' Clinical Medicine. Vol. II.
114. LEXICON of Medical Terms. (Eleventh Part.)

1886. (*Twenty-eighth Year.*)

115. SELECTED Essays on Micro-Parasites in Disease. Edited by W. Watson Cheyne.
116. LEXICON of Medical Terms. (Twelfth Part.)
117. HANDBOOK of Geographical and Historical Pathology. By Dr. Aug. Hirsch. Vol. III.
118. LEXICON of Medical Terms. (Thirteenth Part.)

1887. (*Twenty-ninth Year.*)

119. SPIEGELBERG'S Midwifery. Vol. I. Translated by Dr. J. B. Hurry.
120. LEXICON of Medical Terms. (Fourteenth Part.)
121. SELECTED MONOGRAPHS:— Raynaud's Disturbances of Circulation in the Extremities; Klebs and Tommasi-Crudeli on the Nature of Malaria; Marchiafava and Celli on the Blood in Malaria-Infection; Neugebauer on Spondylolisthesis.
122. ATLAS of Pathology. (Fasciculus VI.)

1888. (*Thirtieth Year.*)

123. SPIEGELBERG'S Midwifery. Vol. II. Translated by Dr. J. B. Hurry.
124. LEXICON of Medical Terms. (Fifteenth Part.)
125. HENOCH'S Diseases of Children. Vol. I.
126. COHNHEIM'S General Pathology. Vol. I.

1889. (*Thirty-first Year.*)

127. ATLAS of Pathology. (Fasciculus VII.)
128. CHARCOT's Clinical Lectures on Diseases of the Nervous System. Vol. III.
129. COHNHEIM's General Pathology. Vol. II.
130. LEXICON of Medical Terms. (Sixteenth Part.)
131. HENOCH's Lectures on Diseases of Children. Vol. II.

1890. (*Thirty-second Year.*)

132. FLÜGGE's Micro-Organisms.
133. COHNHEIM's General Pathology. Vol. III.
134. LEXICON of Medical Terms. (Seventeenth Part.)
135. ATLAS of Pathology. (Fasciculus VIII.) Diseases of Brain and Spinal Cord.

1891. (*Thirty-third Year.*)

136. EWALD's Disorders of Digestion. Translated by Dr. Saundby.
137.
138.

Volumes marked * are now quite out of print.

LIST OF SURPLUS VOLUMES,

With Prices.

N.B.—The prices affixed can be continued only for a limited period until surplus stock is disposed of.

ATLAS OF SKIN DISEASES. Fasciculi 11, 14, 15, 16, and 17. Separately, 10s. 6d. each. Most of the stones have been destroyed, and only a limited number of impressions remain in stock, and a few are out of print.

ON SYPHILIS IN INFANTS. By PAUL DIDAY. Translated by Dr. WHITLEY. 5s.

MEMOIRS ON DIPHTHERIA. By BRETONNEAU, TROUSSEAU, DAVIOT, GUERSANT, BOUCHUT, EMPIS, &c. Selected an Translated by Dr. R. H. SEMPLE. 3s. 6d.

ON THE MINUTE STRUCTURE AND FUNCTIONS OF THE SPINAL CORD. By SCHROEDER VAN DER KOLK.

ON THE MINUTE STRUCTURE AND FUNCTIONS OF THE MEDULLA OBLONGATA, AND ON EPILEPSY. By SCHROEDER VAN DER KOLK. Translated by Dr. W. D. MOORE, of Dublin. In one volume, with numerous Lithographs. 5s.

EXPERIMENTAL RESEARCHES ON THE EFFECTS OF THE LOSS OF BLOOD IN INDUCING CONVULSIONS. By Drs. KUSSMAUL and TENNER. Translated by Dr. BRONNER, of Bradford.

ON THE PROCESS OF REPAIR AFTER RE-SECTION AND EXTIRPATION OF BONES. By Dr. A. WAGNER, of Berlin. Translated by Mr. T. HOLMES. Numerous Woodcuts.

VON GRAEFE'S THREE MEMOIRS ON GLAU-COMA, AND ON IRIDECTOMY AS A MEANS OF TREATMENT. Translated by Mr. T. WINDSOR, of Manchester.
Three Monographs in one Volume. 2s. 6d.

MEMOIRS ON ABDOMINAL TUMOURS AND IN-TUMESCENCE. By Dr. BRIGHT. Reprinted from the 'Guy's Hospital Reports,' with a Preface by Dr. BARLOW. Numerous Woodcuts. 7s. 6d.

A CLINICAL ACCOUNT OF DISEASES OF THE
LIVER. By Professor FRERICHS. Translated by Dr. MURCHISON. Numerous Woodcuts and coloured Lithographs. 2 vols. 12s. 6d. Vol I. separately, 2s. 6d.

CZERMAK ON THE PRACTICAL USES OF THE
LARYNGOSCOPE. Translated by Dr. G. D. GIBB. Numerous Woodcuts.

DUSCH ON THROMBOSIS OF THE CEREBRAL
SINUSES. Translated by Dr. WHITLEY.

SCHROEDER VAN DER KOLK ON ATROPHY OF
THE BRAIN. Translated by Dr. W. D. MOORE, of Dublin. Four Lithographs.

RADICKE'S PAPERS ON THE APPLICATION OF
STATISTICS TO MEDICAL ENQUIRIES. Translated by Dr. BOND.

ESMARCH ON THE USES OF COLD IN SURGICAL
PRACTICE. Translated by Dr. MONTGOMERY.

Five Monographs in one Volume. 5s.

A HAND-BOOK OF THE PRACTICE OF FOR-
ENSIC MEDICINE, BASED UPON PERSONAL EXPERIENCE By J. L. CASPER, M.D., Professor of Forensic Medicine in the University of Berlin. Translated by Dr. G. W. BALFOUR. Vols. II., III., IV 7s. 6d. each.

A GUIDE TO THE QUALITATIVE AND QUAN-
TITATIVE ANALYSIS OF THE URINE. By Dr. C. NEUBAUER and Dr. J. VOGEL. Fourth edition, considerably enlarged. Translated by by W. O. MARKHAM, F.R.C.P.L. With Four Lithographs and numerous Woodcuts. 3s. 6d.

TROUSSEAU'S CLINICAL MEDICINE. Vols. IV.
and V., separately 5s. each. The other vols. are unobtainable, except by new Subscribers.

YEAR-BOOKS OF MEDICINE AND SURGERY.
1859—64. Six Vols. 2s. 6d. each vol.

BIENNIAL RETROSPECT OF MEDICINE AND
SURGERY, 1865—74. 5 vols. 2s. 6d. each.

STRICKER'S MANUAL OF HISTOLOGY. 3 vols.
31s. 6d. Vols. I. and III. separately, 5s. each.

RINDFLEISCH'S PATHOLOGICAL HISTOLOGY.
Vol. II., 5s.

HEBRA'S TREATISE ON DISEASES OF THE SKIN. Vols. 1 to 5, with Index (THE COMPLETE WORK), 21s.

LANCEREAUX'S TREATISE ON SYPHILIS. Two vols. 5s.

NIEMEYER'S LECTURES ON PULMONARY CONSUMPTION. 2s. 6d.

LATHAM'S WORKS. 2 vols. 7s. 6d. Vol. I., 2s. 6d.

LEXICON OF MEDICAL TERMS. Parts I. to V., forming Vol. I; Parts VI. to X., forming Vol. II.; and Parts XI. to XV., forming Vol. III. (in Parts), 21s. each.

GUTTMANN'S HANDBOOK OF PHYSICAL DIAGNOSIS. 5s.

NEALE'S MEDICAL DIGEST. 5s.

GRIESINGER ON MENTAL PATHOLOGY. 10s. 6d.

WUNDERLICH'S MEDICAL THERMOMETRY. 3s. 6d.

COLLES' WORKS. 3s. 6d.

BILLROTH'S CLINICAL SURGERY. 7s. 6d.

KOCH ON WOUND INFECTION. 2s. 6d.

Several of these works are well suited for presents to Students or for Class Prizes. Amongst them may be especially mentioned STRICKER's Histology; FRERICHS On Diseases of the Liver; LATHAM's Works; HEBRA's Diseases of the Skin; and GUTTMANN's Physical Diagnosis, &c.

LAWS OF THE NEW SYDENHAM SOCIETY.

I.—The Society is instituted for the purpose of supplying certain acknowledged deficiencies in the existing means of diffusing medical literature, and shall be called "THE NEW SYDENHAM SOCIETY."

II.—The Society shall carry out its objects by a succession of publications, of which the following shall be the chief:—1. Translations of Foreign Works, Papers, and Essays of merit, to be reproduced as early as practicable after their original issue. 2. British Works, Papers, Lectures, &c., which, whilst of great value, have become from any cause difficult to be obtained, excluding those of living authors. 3. Annual Volumes consisting of Reports in Abstract of the progress of the different branches of Medical and Surgical Science during the year. 4. Dictionaries of Medical Bibliography and Biography. Those included under Nos. 1 and 2 shall be held to have the first claim on the attention of the Society; and the carrying out of those under Nos. 3 and 4 shall be considered dependent upon the amount of funds which may be placed at its disposal.

III.—The Subscription constituting a Member shall be One Guinea, to be paid *in advance* on the 1st of January annually, and it shall entitle the subscriber to a copy of every work published for that year. *No books shall be issued to any Member until his subscription for the year has been paid.*

IV.—The Officers of the Society shall be elected from the Members, and shall consist of a President, sixteen Vice-Presidents, a Treasurer, a Secretary, and a Council of thirty-two, in whom the power of framing Bye-laws and of directing the affairs of the Society shall be vested. Twelve of the Council shall be provincial residents.

V.—Five Members of the Council shall form a quorum.

VI.—The Officers of the Society shall be elected by ballot at the General Anniversary Meeting of the Society. Balloting lists of Officers proposed by the Council, with blank places for such alterations as any Member may wish to make, shall be laid on the Society's table for the use of Members.

VII.—The President, Vice-Presidents, and Council, shall be eligible for re-election, except that of the Vice-Presidents four, and of the Council eight, shall retire every year.

VIII.—The Council shall appoint local Honorary Secretaries wherever they shall see fit.

IX.—The business of the President shall be to preside at the Annual and Extraordinary Meetings of the Society; in his absence one of the Vice-Presidents, or the Treasurer, or any Member of the Council chosen by the Members present, shall take the Chair.

X.—The Treasurer, or some person appointed by him, shall receive all moneys due to the Society.

XI.—The money in the hands of the Treasurer, which shall not be immediately required for the uses of the Society, shall be vested in such speedily available securities as shall be approved by the Council.

XII.—The Council shall select the Works to be published by the Society, and shall make all arrangements, pecuniary or otherwise, in regard to their publication. In the event of any Member of the Council being appointed to edit any Work for the Society, for which he is to receive pecuniary remuneration, he shall immediately cease to be a Member of the Council, and shall not be eligible for re-election till after the publication of the Work.

XIII.—The Council shall lay before the Members at each Anniversary Meeting a Report of their proceedings during the past year, and also an account of the Receipts and Expenditure of the Society; and shall further cause to be printed and circulated among the Members an abstract of such Report and Accounts immediately after such Anniversary Meeting.

XIV.—The annual Accounts of the Receipts and Expenditure of the Society shall be audited by a Committee of three Members, selected at the preceding Anniversary Meeting from among the Members at large.

XV.—The Secretary shall have the management of the general Correspondence of the Society, and of such other business as may arise in carrying out its objects.

XVI.—The local Secretaries shall further the objects of the Society in their respective districts, and shall be in communication with the metropolitan Secretary.

XVII.—The Anniversary Meeting shall be held in the same town as, and at the time of, the Annual Meeting of the British Medical Association, notice of it having been given to all Members at least a week before the day fixed on.

XVIII.—The Members generally shall be invited and encouraged to propose Works, &c., and to make any suggestions to the Council they may think likely to be useful.

XIX.—The Works of the Society shall be printed for the Members only.

XX.—No alteration in the Laws of the Society shall be made, except at a General Meeting. Notice of the alteration to be proposed must also have been laid before the Council at least a month previously.

XXI.—The Council shall have power to call a General Meeting of the Members at any time, and shall also be required to do so within three weeks, upon receiving a requisition in writing to that effect from not less than twenty Members of the Society.

XXII.—All Special General Meetings of the Society shall be held at such place as the Council may appoint.

XXIII.—The Council shall meet at least once in two months, unless by special resolution to the contrary.

GENERAL INFORMATION.

A THIRD EDITION of the VOLUMES for 1859 was printed, and a second of that for 1860. For subsequent years the First Edition was much larger; and it is not likely that any of the Volumes will be reprinted.

Most of the stones for plates, &c., both those for the Atlas of Skin Diseases and those for printed Volumes, have been destroyed, and will not be reproduced.

The Society is now in its Thirty-third year. Arrangements have been made by which new Members can obtain single Volumes, or sets of Volumes, from the Society's stock in hand. Some of the Volumes, of which a larger surplus exists than of others, can be purchased at fixed prices (for which see list). The Society's Agent is empowered to make special arrangements with new Members who may wish to obtain any of the past Volumes.

CARRIAGE, &c.—The Society's Works are supplied free of cost to any address in London, Edinburgh, or Dublin; but the expenses of Carriage to all other places must be borne by the members to whom they are sent. Members are requested to give detailed instructions respecting the mode by which they wish their Volumes to be forwarded, and also to remember that the Society's responsibility ceases when the Book has been delivered according to the instructions given. Members residing in the British Isles wishing to receive their Works by post can do so by prepaying the sum of 2s. for the year for postage.

The Subscription is One Guinea annually, to be paid IN ADVANCE. The best mode of sending money is by Cheque, Post-office or Postal Order, payable to Mr. H. K. LEWIS; or by Cheque to the order of the Treasurer, Dr. SEDGWICK SAUNDERS. It is requested that in future all communications in reference to the payment of Subscriptions, or the issue of Books, may be made to Mr. LEWIS, the Society's Agent, and not to the Secretary.

JONATHAN HUTCHINSON,
Hon. Secretary.

15, CAVENDISH SQUARE, W.

N.B. — To prevent misapprehensions as regards the punctual issue of each year's series it seems desirable to reprint the following extract from the Report for the year 1882:—

"If the members would kindly understand that the Society's financial year is from January to December, its year of issue from June to June, and that its subscriptions are due in advance, the working of the Society would be much facilitated. From this point of view, the issue of volumes for each succeeding year has always in the past been punctually completed, and probably will be so in the future. The works promised are issued *for* the year specified, but are not all of them issued *in* it."

⁎ Any Member wishing for additional Copies of this Report, &c., can obtain them by applying to Mr. HUTCHINSON; or to the Society's Agent, Mr. LEWIS, 136, Gower Street, W.C. The Council will be much obliged by its distribution amongst those thought likely to join the Society.

IMPORTANT NOTICE TO NEW SUBSCRIBERS AND LOCAL SECRETARIES.

New Members who subscribe for the current year and not fewer than three past years at the same time will be allowed to select volumes from the surplus stock to the value of one guinea without additional payment. The like privilege will be secured each year by any Local Secretary who has the subscriptions of all the members on his list (the number being not less than ten) paid before the end of March for the current year.

PS.—The Society's Agent is prepared to supply, at fixed prices, CASES for binding the Lexicon, and PORTFOLIOS for the reception of the Plates of Skin Diseases, and for the Pathological Atlas.

Hon. Secretary.

JONATHAN HUTCHINSON, Esq., F.R.S., 15, Cavendish Square, London, W.

Agent and Depôt for Books.

Mr. H. K. LEWIS, 136, Gower Street, London, W.C.

WEST, NEWMAN AND CO., PRINTERS, 54, HATTON GARDEN, LONDON, E.C.